Explaining the Solar System

Student Exercises and Teacher Guide for

Grade Nine Academic Science

Jim Ross   *The University of Western Ontario*

Mike Lattner   *Algonquin and Lakeshore Catholic District School Board*

rosslattner educational consultants   *London Ontario Canada*

National Library of Canada

Cataloguing in Publication

Ross, Jim (James William), 1952-
Explaining the solar system : student exercises and teacher guide for grade nine academic science / Jim Ross, Mike Lattner.

1. Solar system--Study and teaching (Secondary)--Activity programs.

2. Solar system--Study and teaching (Secondary)
    I. Lattner, Mike, 1957-
    II. Title.

QB501.R68 2003   523.2'071'2   C2003-901189-0

Authors

Mike Lattner
Jim Ross

Printer

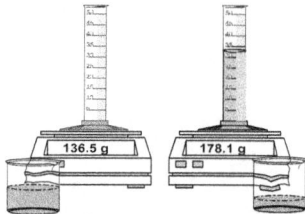

ISBN   978-1-897007-03-7

Offices   Wellington Ontario Canada
              London Ontario Canada

To teachers, parents and students everywhere who desire to bring about new ways of understanding the world.

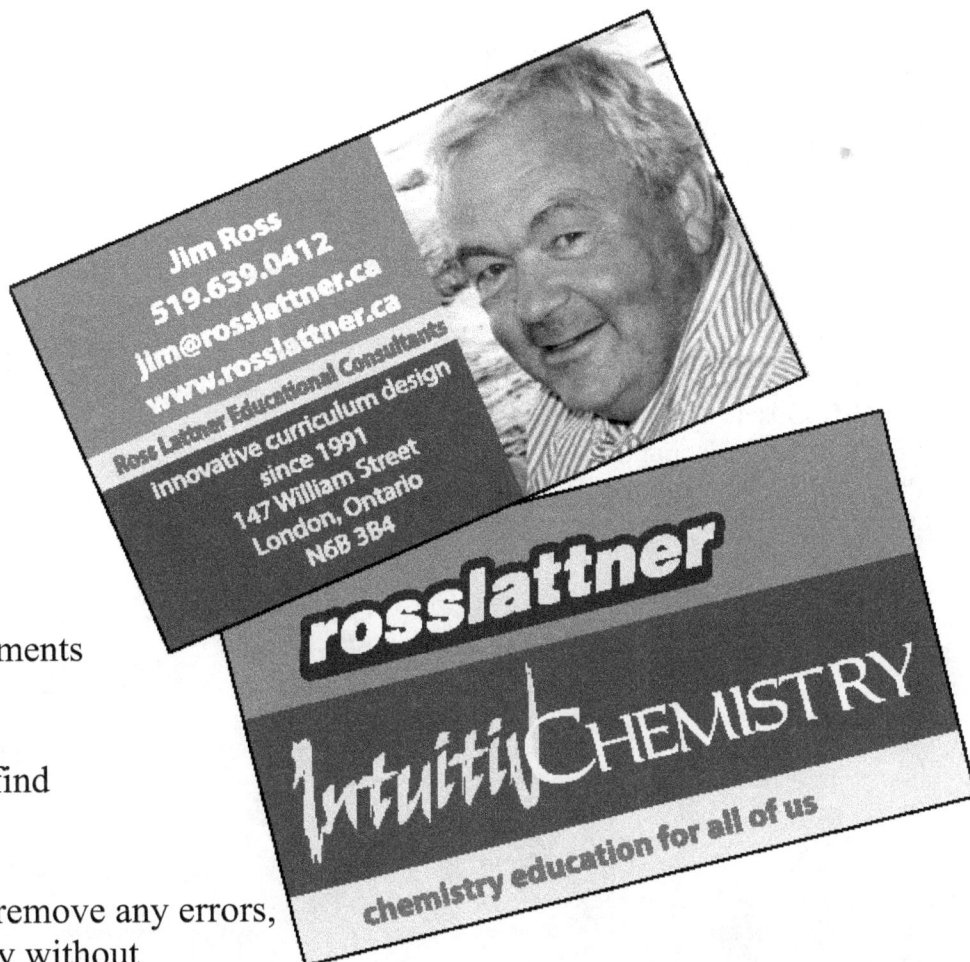

Jim Ross
519.639.0412
jim@rosslattner.ca
www.rosslattner.ca

Ross Lattner Educational Consultants
innovative curriculum design
since 1991
147 William Street
London, Ontario
N6B 3B4

rosslattner
Intuitiv CHEMISTRY
chemistry education for all of us

We welcome your comments and suggestions.

Let us know what you find most useful.

We've worked hard to remove any errors, but don't let a day go by without letting us know if you find one.

Stay in touch.

Jim Ross

Our thanks to all of the wonderful people at the Faculty of Education, the University of Western Ontario.

Special thanks to Robert Duff, always willing to talk to us about astronomy.

Explaining the Solar System

Table of Contents

Explaining the Solar System

Table of Contents

Explaining the Solar System

1 Teaching About the Solar System

Title:	Explaining the Solar System
Time Allocation:	27.5 hours (22 periods of 75 minutes each)
Authors:	Jim Ross and Mike Lattner
Date:	May 2003
Unit Description:	This unit is a highly descriptive account of the contents and history of the universe, as it is pictured by scientists over the past fifty years. The four forces that shape our universe are emphasized as explanatory tools. Two major projects are used to cover the vast number of topics and resources available on this topic.

The unit itself is subdivided into four major sections. Each section will take a little more than one week to complete.

1. Construction of some simple instruments to assist in observation of the earth, the sun, the moon, and time, all from the point of view of the surface of a tilted and spinning earth.
2. Observation of the sun, the closest star. From these observations we move to an understanding of the forces that shape our sun and its behaviour.
3. Observation of the moon, and other satellites. From these observations we build an understanding of the role of gravitation in orbital motion
4. Research project on phenomena within our solar system
5. Research project on phenomena beyond our solar system

This time line is divided into separate one-week blocks. Students could observe and record several lunar cycles, or the changing position of the sun, over the entire semester.

Strand:	Earth and Space Science
Expectations:	Overall Expectations: ESV 1 - 3 Specific Expectations: ES1.01-1.07, 2.01-2.03; 3.01-3.03

Explaining the Solar System

How do young people sort out the science from the science fiction from the science legend?

Good *science* is explanatory, and capable of prediction.

Good *science fiction*, like good science, starts with a shared understanding of really strong theories, then develops and extends those theories into a human situation.

Science legends are stories that are only minimally connected to everyday experience and to explanation. They have no apparent purpose. They just *are*.

Much student talk about the universe is legend, not science!! Listen carefully to your students!

Unit Planning Notes:
This unit explores some of the features of the visible universe. The intended emphasis is upon observable phenomena which exemplify the four forces which shape our universe.

Some time is spent upon the problem of making observations of things in the night sky. Some of these objects have observable movements of their own. All of these objects are observed from a tilted, spinning, revolving Earth. Sorting out any real motions from the apparent motions is one difficulty that students face. A more challenging problem is the student construction of a "sense of location" in space that permits a student to anticipate the appearance of bodies such as the sun, the moon, the stars, and the planets, based upon previous observations.

The second half of the unit is devoted to a pair of projects which explore points of interest within the solar system, followed by the exotica of the rest of the universe.

Prior Knowledge Required This unit assumes that the student has heard a large number of the terms used in this unit, and has likely used them personally. The student probably does not have an extensive mental catalogue of objects and forces, but is familiar with light and with gravity. While a typical student does not have a strong, theory-based understanding of the objects in the universe, he or she does have some degree of commitment to a scientific world view.

The only way that the scientific legends that surround the study of the universe can be addressed is by calling upon the student's commitment to coherence, evidence, and explanation.

Teaching and Learning Strategies The focus of science is not nature itself. The focus of science is our shared *representations* of nature. Accordingly, three learning strategies are emphasized. Students are expected to commit themselves to a *prediction* of the behaviour of each demonstration or lab. Students are expected to explain why they believe their prediction, in both *pictorial representations* and in *sentences*. Students are expected to gradually master a small set of *theoretical propositions*, and then to increasingly represent their arguments in terms of the theory.

Assessment and Evaluation A variety of strategies and instruments will be used throughout this document. Projects are a major component in this unit.

The student learning goals of this unit are: To predict the location and appearance of the sun, moon and stars against the sky; To explain the "whys" of their appearance.

Four structurally simple ideas that scientists used to explain everything in the universe.

Should the "Gee Whiz" legends of the universe be permitted to distract us from our terrestrial home base?

Wonder.....

What is it really like "out there?" Where did we come from? Are we alone? How did it all begin? How will it all end? Human beings have asked these questions from the very beginning of our existence, and will continue to ask them into the distant future. Some of these questions may never be answered. This unit consists of three main ideas:

1. **There are only Four Forces** In all of the material universe, four forces connect everything that we can perceive. These forces are the same everywhere: gravity near a black hole is basically the same as gravity on earth, only more intense. We are a part of the universe, so those four forces are present in our bodies and our neighbourhoods as well as the distant galaxies.
    1. **Gravity: a very weak force of attraction between any two things with mass**
    2. **Weak Nuclear Force: a strong force that governs radioactivity in the nucleus**
    3. **Electric Force: a strong force between any two things with charge. Light is the carrier of the electric force.**
    4. **Strong Nuclear Force: an extremely strong force that holds nuclei together**

    Sun, moon and stars, day and night, summer and winter, the passage of years, the stuff you are made from, the planets in our solar system, absolutely everything involves those four forces.

2. **There is only One Earth** upon which we observe the universe. Because the earth we stand upon is spinning, tilting, and revolving around the sun as we look at the rest of the universe, our observations are made difficult. Some of this course will help us to make sense of the apparent motions of the sun, the moon, the planets and the stars as we spin, tilt and revolve with our earth. One of the most humbling things about observing the night sky is the size of it all, and how small we are against it.
    Our study of space must always reflect the finiteness of our own planet, and the precious ways in which the four forces have shaped our planet to support our lives.

3. **Where do we go from here?** For some of us, our journey as human beings might be a physical voyage to the moon or other planets. Others may study and work to take a journey of exploration, using telescopes, space probes, the Internet and other technological devices. For all of us, a journey of understanding is always possible. We will study the structure of the observed universe, from the most local to the most remote, and from the present to the distant past.

Explaining the Solar System

Project 1.1: Build and Use an Astronomical Instrument

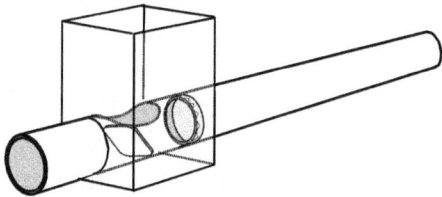

Pedagogical Issues Two issues are to be addressed here. First, how do we make students aware of the processes that make our sun work? Second, since we wish to observe the sun, how does a student learn to take into account the apparent motion of the sun?

A pinhole telescope should provide simple images of the sun. The motion of the earth should be readily apparent, and student solutions to that problem may be encouraged to emerge. It might be possible to observe very large sunspots.

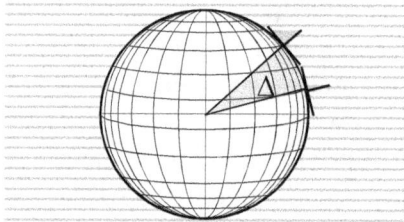

Science Issues Four construction projects are proposed: a pinhole solar telescope, a measurement of earth's diameter, a sundial, and a zodiac globe. These projects are concerned with building instruments to observe the apparent behaviour of the sun, *as seen from a tilted, spinning, revolving earth.*

All projects involve the use of saws and drills. Students should do this work under the supervision of an adult, preferably a parent.

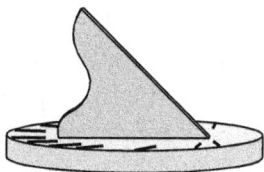

The pinhole solar telescope should provide some of the basic understanding of the difficulties encountered when trying to point an earth-bound telescope at the stationary stars. In addition, it will provide a way to measure the size of the sun.

With the internet, an atlas, and a couple of long-distance telephone calls, you can measure the size of the earth.

A sundial is a beautiful work of art linking earth and sun.

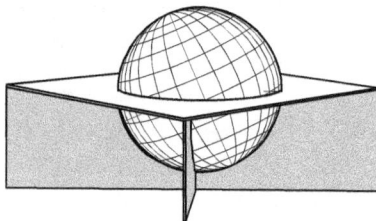

The zodiac globe helps students to understand the seasonal changes in the appearance of the night sky.

The Learning Activity

If this activity is introduced on a Monday, students will be able to finish the project either at home or at school in five days plus the weekend.

Your planning will depend upon your students' choice of workplace. If they choose to work primarily at home, then you can undertake other activities during the class time. You will not be able to assign homework. Alternatively, you can attempt to do the construction at school, and assign short observations for homework.

Equipment, Preparation and Resources

Heavy cardboard tubes

Light plywood, paneling, or masonite

Hand saws, Hand drills, Keyhole saw, Screwdrivers

Measuring tapes, Metre sticks, Straight edges, Protractors

White glue, clamps, elastics, tape, string

White plastic grocery bags

Some globes, or a basketball and acrylic paint

Paint brushes, leftover latex paint

Categories:	Assessment and Evaluation
Knowledge:	student knowledge of design and use of instrument
Inquiry:	Various potentials exist
Communication:	Written report
Applications, Extensions:	History, lore of instruments

Lab 2.1: What Makes a Star Work?

Consider first a star like our sun. To burn all of its hydrogen fuel takes about ten billion years.

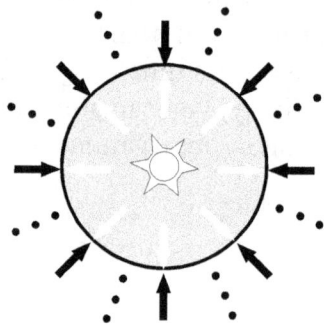

Now consider a star with ten times the mass. Greater gravity compresses the hydrogen together, so that the pressure and temperature in the core is much greater. Its nuclear reactions will consume its tenfold greater mass in only one billion years.

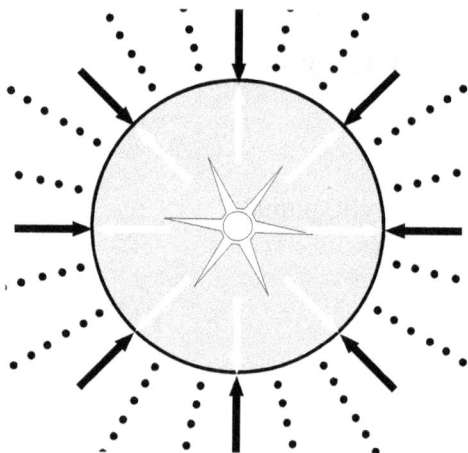

Pedagogical Issues It would be possible to make an elaborate static model of the sun at this point, and require the students to "learn" it. This is sometimes frustrating for students. Static models often lack the dynamic causative structures which play such an important cognitive role for students.

"Telling the story" of a star requires the student to possess some cognitive model of the forces involved.

Science Issues The basic dynamics of the sun require only two forces: gravity, and the electromagnetic force. Gravity pulls the matter of the star inward, increasing the pressure and temperature. Light is the carrier of the electromagnetic force. Light supplies the outward force which keeps the sun from collapsing.

Suppose that two stars are "burning" hydrogen fuel in a nuclear fusion reaction. The second star is ten times the mass of the first. Students usually do not find it difficult to understand that the more massive star is subject to stronger gravitational forces. Yet the star does not collapse... Something must be able to resist this greater gravitational force. This something is a greater outward force of light, both more copious and more energetic. Large stars are both brighter and bluer than smaller stars, the composition of the stars being equal.

Of course, one might ask why this extra light is being produced. Under the greater gravitational confinement in the core of the larger star, the fusion of hydrogen into helium proceeds at a much greater rate. A dynamic steady state is achieved, in which greater gravity causes greater rates of reaction, which produces greater light. The star's gravity collapses the star to the point that the pressure of light balances gravity.

A star like our sun will eventually spend all of its hydrogen fuel. Without a nuclear reaction, it gradually collapses under gravity, increasing the core T and P until a new set of nuclear reactions starts.

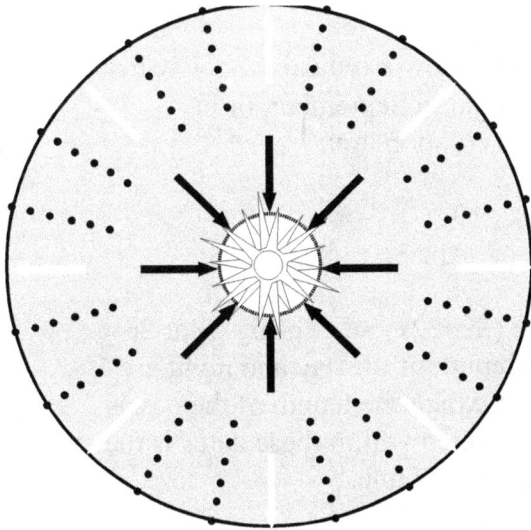

Helium is fused into carbon, oxygen, and nitrogen. This reaction continues in a tiny, hot core, but the light energy pushes the envelope of the sun out to the orbit of Mars.

The Learning Activity

This simple model is an example of a kind of "steady state" in which the inward tension of the rubber skin of the balloon is balanced by the outward pressure of the air.

Adding air to the balloon is analogous to increasing the mass of a star. Both the inward tension of the rubber skin and the outward pressure of the air are increased.

Before the experiment
Predict: What happens to the gravitational force? The interior pressure? The interior temperatures? the rate of the fusion reaction? The brightness of the star?
Explain: why you believe your predictions.

Predict: what happens to a star when it uses up its hydrogen fuel?
Explain: why do you believe your predictions?

What happens next is complex, and depends entirely upon the initial mass of the star. Stars can fizzle. They can start fusing helium nuclei, or even heavier elements. Stars can shrink, swell, or collapse to a point. They can fade, go nova, or super-nova. They can end as white dwarfs, red giants, or black holes. All of these fates depend only upon the initial mass of the star, and the four forces.

Equipment, Preparation and Resources
Balloons

Categories:
Knowledge:
Inquiry:
Communication:
Applications, Extensions:

Assessment and Evaluation
describes relationship between mass, brightness, and colour
develops "what-if" questions
quality of written explanation.

Explaining the Solar System

It's ten o'clock at night. Can you point to the position in the sky that the sun will occupy at ten o'clock tomorrow morning?

The apparent path of the sun is also, within 5°, the apparent path of the moon and the planets. The sun, the moon, and the planets, however, all follow different rhythms.

Any application of solar energy on this planet requires familiarity with the apparent path of the sun.

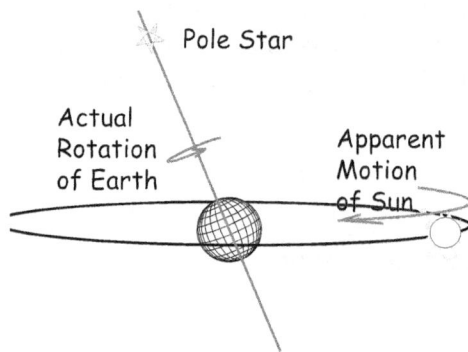

Lab 2.2: Observing Sunrise, Sunset and Elevation

Pedagogical Issues All of our everyday language speaks of the sun going around the earth. Outside your classroom, there is probably not one compelling reason for your students to adopt a heliocentric model of the solar system. This activity is designed to at least familiarize the students with the apparent motion of the sun, so that they are prepared to predict the position of the sun, on the basis of extended observations.

Science Issues Assuming we follow a typical school year, observations on this unit could begin in September, or in February. Suggested dates of observation are:

> Sept 7, 14, 21, 28, Oct 5, 12, 19   or
> Feb 19, 26, Mar 6, 13, 20, 27, Apr 3,

These dates include the Autumn (Sept 21) and Spring (Mar 20) Equinox, the dates at which the length of the day and night are equal. These are also the dates at which the length of the day is changing most rapidly. Also bracketed within these dates is the change of clocks for Daylight Savings Time.

The apparent path of the sun lies along the Ecliptic, roughly the plane of solar system. The moon and the planets are confined to this plane, and appear to follow the same path. We will be observing them in later activities.

The apparent location of the ecliptic changes with the seasons, due to the tilt of the earth's axis, 23.5°. It is highest in the sky on Summer Solstice (June 21), when it is "*Latitude + 23°*", and lowest on Winter Solstice (Dec 21) when it is "*Latitude − 23°*."

Observing the Sun

A concept which cannot be "seen at a glance" is unlikely to be developed by a student without assistance.

Many students have heard that the earth goes around the sun. Very few are able to point to concrete evidence to support this idea.

This exercise requires students to organize a relatively large collection of representations into a single explanatory framework.

The Learning Activity This is a very long-term activity, extended over at least six weeks, and perhaps as much as four months. It is important to keep checking on students, making sure that they either observe sunrises and sunsets themselves, or failing that, to check the newspaper for the information.

This diagram shows the path of the sun, following the ecliptic, at two days during one semester. The apparent path of the moon and the planets is also very near the ecliptic.

Equipment, Preparation and Resources Students must select a place from which to view the sun. A park, hilltop, apartment window may be suitable.

Newspapers often print sunrise, sunset and other information in the weather section.

Some students may have difficulty locating due south. Compasses and /or maps may be helpful.

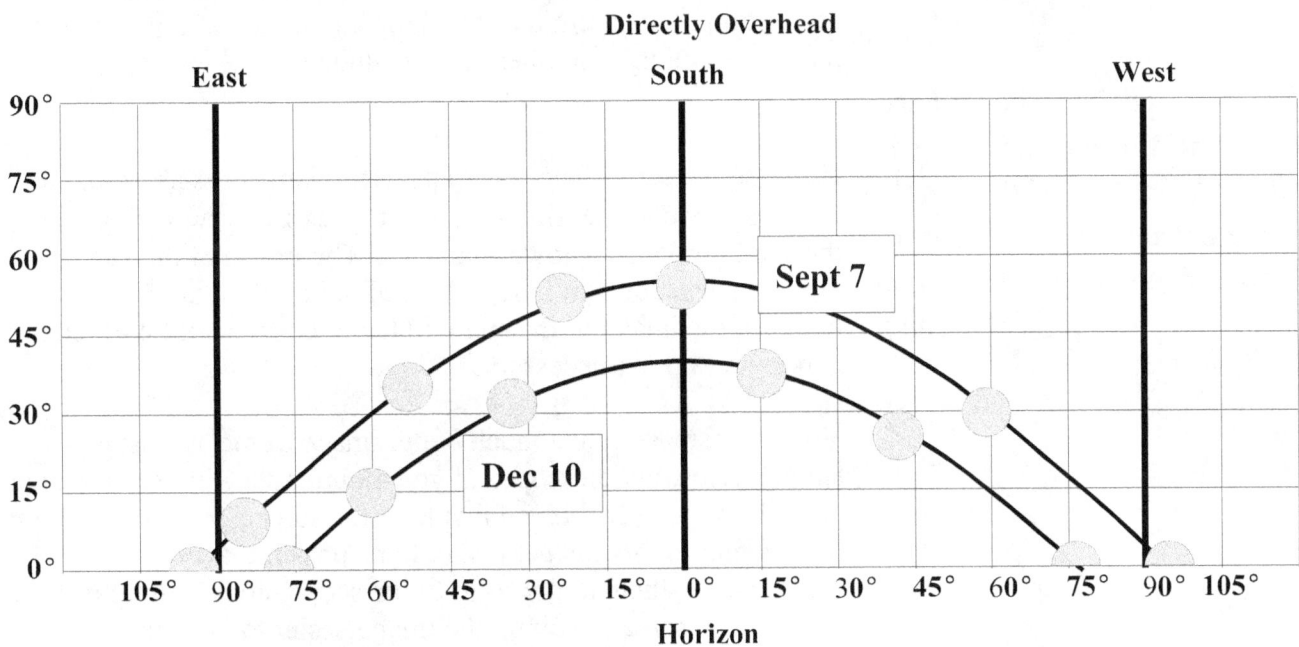

Categories:
	Assessment and Evaluation
Knowledge:	Displays knowledge growth over an extended period of time
Inquiry:	Keeps records over time
Communication:	Clarity of diagrams and records
Applications, Extensions:	Explains the ideas behind daylight savings time

Explaining the Solar System

Lab 2.3: Observing the Sun with a Pinhole Telescope

Viewing an image of the sun in a pinhole telescope. Aim the scope by positioning it so that its shadow is minimized. The scope must be mounted on some chairs with tape.

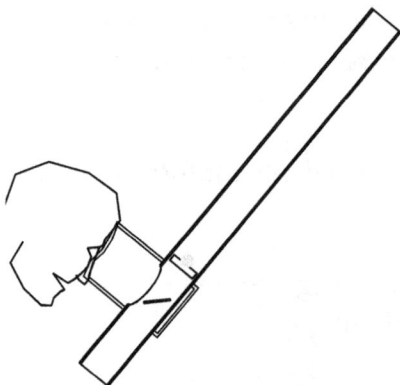

Pedagogical Issues The first set of issues arise from the operation of a pinhole telescope. The brightest image is obtained by using a relatively large pinhole. With a larger pinhole, students are more likely to locate the image of the sun, and fix the telescope to observe it for a few minutes. Unfortunately, a large pinhole means a fuzzy or blurry image.

After successfully viewing the image through a large pinhole, some students may wish to use a small pinhole for a dimmer, but clearer, image.

The second set of issues arise from the difficulty of fixing upon the image of the sun for a length of time. Because of the earth's rotation, the image moves across the screen. In 10 minutes, it may move out of view, necessitating a telescope adjustment. Usually, people are not aware of the motion of the sun in the middle of the sky, because there is no point of reference. The perception of apparent motion along a circular path may provoke questions among a number of your students.

The image produced on a card by the terrestrial scope is much brighter. Cut a shield for the front of the scope to reduce the sunlight falling on the card.

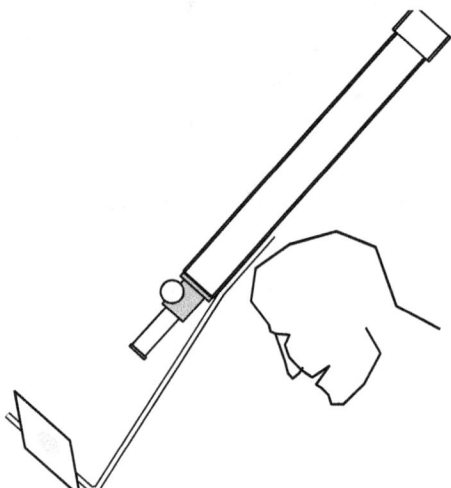

Science Issues The sun's apparent diameter is about 0.5 degrees of arc, rather tiny in fact. It takes about two minutes for the sun to traverse its own diameter. The image of the sun in a pinhole telescope 1 m long will be about 8 - 9 mm. In 10 minutes, the image of the sun will have moved about 4 - 5 cm, making re-aiming necessary.

You can demonstrate a much better image of the sun using a simple terrestrial telescope. If you mount a viewing screen 20 cm back from the eyepiece, an image of the sun can be projected onto the screen. In this image, it should be possible to observe sunspots. A shield at the front of the scope blocks sunlight from the viewing screen, making the image easier to observe.

Observing the Sun

The Learning Activity   There are several objectives.

What other possibilities exist for exercises of this type?

First, simply to produce an image of the sun. To produce an image in the pinhole telescope requires patience and a degree of skill.

Second, to observe the apparent motion of the sun, magnified in the pinhole telescope.

Third, to examine the image itself for the presence of sunspots, etc.

Some senior students may be familiar with video capture boards. With an inexpensive camera, some filters, cable, etc, observation could be automated.

Let the students achieve some success in this activity before pre-empting their achievement with the much superior image produced by a terrestrial telescope and mounted card.

In the activity following this one, the students are to measure the size of the image, and calculate the size of the sun. Perhaps the students have refined their observations to the extent that they can make a measurement today, and simply perform the calculations tomorrow.

If the school has a darkroom, perhaps one of the telescopes can be fitted with some photographic paper, and sealed up in the darkroom, making a pinhole camera. When the camera is aimed, the pinhole is opened for a few seconds to make an exposure on the paper. Developing the paper will produce a permanent image which can be measured later.

Equipment, Preparation and Resources

Pinhole telescopes

Chairs, masking tape

Terrestrial telescope

Categories:

Knowledge:

Inquiry:

Communication:

Applications, Extensions:

Assessment and Evaluation

Describes how the image of the sun forms in a pinhole camera

solves practical problems related to obtaining an image

Clarity of diagrams and records

Lab 2.4: How Big is the Sun?

Do students readily transfer knowledge and procedures from one area to another?

A student may be able to use geometric knowledge (similar triangles) and arithmetic knowledge (ratios) to answer questions in a math text.

The same student may not be able to calculate the size of the sun as described in the lab exercise.

Don't believe it?

Let the students struggle with this problem for a few minutes. Many of them may need a "bridging device," perhaps a large diagram, a loop of string, some metre sticks, etc.

Pedagogical Issues   A student's sense of distance and place is based upon his or her own experience in the body. A distance related to touching or a brief walk is readily understood, but the distance to a point out of the field of vision is not well understood, even by many secondary school students. The student will readily understand the dimensions of the phenomenon within the telescope, has no experiential point of reference for the astronomical distances.

The cognitive problem is the utter abstraction of a number like 150 000 000 km, the distance to the sun.

Science Issues

The diameter of the earth is 12 800 km, or let's say 15 000 km for the sake of a nice ratio. If the earth was a 1 cm ball, a marble perhaps, the sun would be 10 000 times earth's diameter away, or 100 m away. The moon would be 30 cm away, about the size of a dried pea. The sun would be 1 m in diameter.

Since the students may be outdoors for this exercise, perhaps some very long strings could be used to model a pair of similar triangles with enormously different relative sizes.

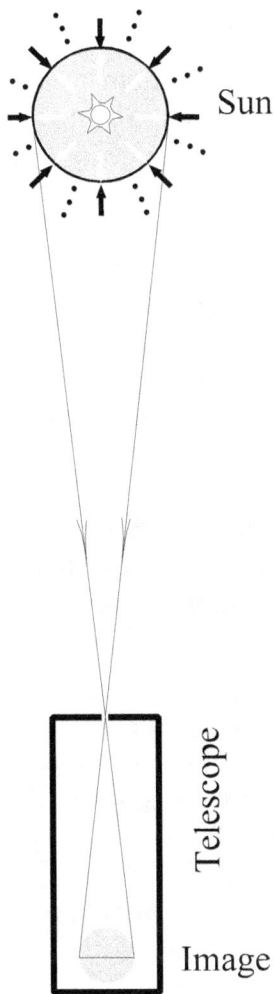

The Learning Activity

Before the experiment

    Predict: the size of the sun, in metres, without consulting an authority.

    Explain: your prediction

The students aim one or more pinhole telescopes at the sun, and measure both the size of the image and the length of the telescope.

After the experiment

    Observe: the size of the image, and calculate the size of the sun if it is 150 000 000 000 m away.

    Explain: any differences between your prediction and observation.

The calculation is likely to be the most challenging aspect of this activity.

Sun

Telescope

Image

Equipment, Preparation and Resources

Pinhole telescope
Pencil and paper
Chairs for supports
Masking tape, etc.

Categories:

Knowledge:

Inquiry:

Communication:

Applications, Extensions:

Assessment and Evaluation

Describes how the image of the sun forms in a pinhole camera

solves practical problems related to obtaining an image

Clarity of diagrams and records

Explaining the Solar System

Lab 3.1: Observing Earth's Largest Satellite

Of course the moon goes around the earth. That's "common sense."

But "common sense" often is not strongly related to scientific explanation. The apparent motion of the moon on a given night, from moonrise to moonset, is *not*, by itself, evidence that the moon goes around the earth!!

If students keep an accurate set of records over an extended period of time, they may be able to integrate those records into a coherent explanation of the moon's changing position and appearance.

Simple as it looks, this is not an easy task. The everyday sense that we stand upon a stationary ground is awfully compelling!

Moonrise is about one hour later each evening. Can your students explain why?

Pedagogical Issues This activity is similar in many ways to *Lab 4.1.2: Observing Sunrise, Sunset and Elevation.* In fact, students are likely to enter this activity as if the phenomenon of the moon's cycle was identical to that of the sun.

Consider yourself fortunate if they do. The most intractable learning obstacle that students face is the completely understandable idea that they reside on a stationary platform, and the moon goes around us. Therefore, the apparent motion of the moon during the course of a single evening will be taken as evidence that the moon is in orbit around the earth.

It is from records of the apparent position of the moon *at the same time each day over at least twenty eight days* that students may be able to comprehend the orbital motion of the moon around the earth.

Science Issues Over the time interval of one evening, the apparent motion of the moon from east to west is in fact due to the rotation of the earth from west to east.

Over the time interval of one month, the slow progression of the moon from west to east is due to the proper motion of the moon in its orbit around the earth.

In fact, the moon can be seen to pass in front of a star from time to time. When that happens, the moon occultation appears on the Eastern side of the moon. In other words, as the moon appears to be moving east to west due to the rotation of the earth, it is ever so slowly moving west to east in its own orbit.

The Motion of Satellites

Each student's observations must be made at the same time each night!!

Jennifer observes after her homework at 10:00 PM

Mark observes on his way to practice at 7:00 AM

Lisa observes on her way home at 5:00 PM

Adam observes during his lunch period at 12:00 noon.

The Learning Activity This observation will take at least one full lunar cycle, about 28 days. Students become better informed about the motion, and hence better observers, if the observations extend over two or more cycles. You may even try to have them predict the position of the moon on a given date in the near future.

This diagram shows the apparent path of the moon over a fourteen day period. The actual apparent path can vary considerably, because the moon's orbit is tilted about 5° w.r.t. the earth's orbital plane.

Equipment, Preparation and Resources Students must select a place from which to view the moon. A park, hilltop, apartment window may be suitable. A recording system must be developed. The shape and position of the moon must be recorded, as in the diagram below:

Directly Overhead   9:00 PM Each Evening

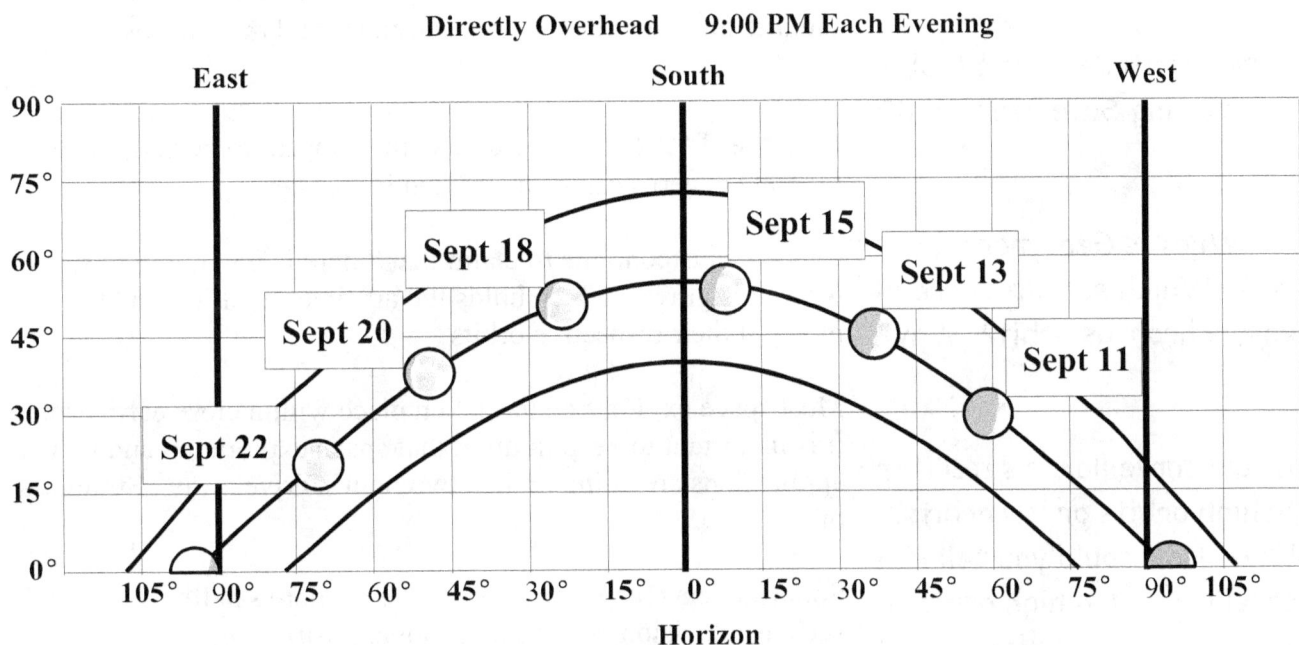

Categories: **Assessment and Evaluation**
Knowledge: Displays knowledge growth over an extended period of time
Inquiry: Keeps records over time
Communication: Clarity of diagrams and records
Applications, Extensions: Predicts a date of full moon at least two months in advance.

Lab 3.2: Observing Earth's Smallest Satellites

Can students develop a taxonomy of artificial satellites?

Pedagogical Issues What do your students think a *satellite* is? Ask them to write their ideas down on a piece of paper, before they have a chance to talk together. You may find more holes and gaps in their knowledge than you might have imagined.

Orbit direction: North to South, South to North, East to West, West to East. Which kinds are most commonly observed? Which least observed? Why?

Students tend to think of satellites as things that do something (take pictures, transmit television shows), without having a clear idea what they actually are. Most students are not able to make connections between the moon and artificial satellites as being similar in their orbits around earth.

The concept of Gravity, the weakest fundamental force, is the concept that unifies the diverse behaviours of artificial satellites and the moon.

Students may draw parallels between the behaviour of a tiny satellite of the earth, and an electron around the nucleus.

At what altitude to the Global Positioning Satellites orbit?

Science Issues This exercise may appear to be peripheral, but several concerns can be addressed in this lab.

What is Geosynchrohous orbit? Which satellites are in Geosynchronous orbit? Why?

Can students continue to claim that "there is no gravity in space" when it is gravity which holds the artificial satellites and the moon in their respective orbits?

Most space exploration has taken place within close earth orbit. It is important to keep in mind that space exploration and related applications are going on just two hour's drive away. Straight up, of course....

Would surveillance satellites be in high orbits or low orbits? Why? How could you tell if a satellite was in high orbit or low orbit?

Some students may subscribe to one or more satellite services, such as television reception or global positioning.

The Motion of Satellites

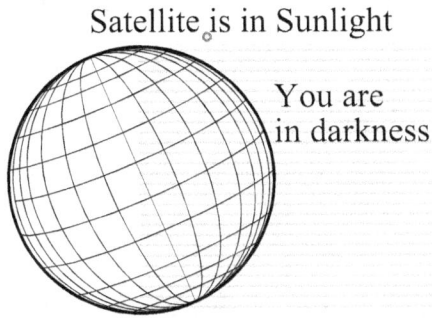

The Learning Activity requires about half an hour of night-time viewing., as soon as the sky is black. The satellites must be far enough above the earth to be in sunlight, even as the surface of the earth is in darkness. Satellites are rarely observed later than two hours after sunset.

Satellite is in Sunlight

You are in darkness

Before the experiment
    Predict: how many satellites does the student expect to see
    Explain: the prediction.

After the experiment
    Observe: satellites for a 30 minute period.
    Explain: observations.

Equipment, Preparation and Resources

Naked eye viewing. A pair of binoculars may help to see other objects of interest, but it's pretty hard to catch a satellite in high powered binoculars.

Categories: **Assessment and Evaluation**
Knowledge:
Inquiry: Quality of records kept for the observations
Communication:
Applications, Extensions: Expresses knowledge of applications of satellites

Lab 3.3: The Sun and Moon's Gravitational Effects Upon the Earth

The imaginative terms " black hole", "stellar nursery", "red giant" and so on sit firmly in the neighbourhood of *science legend*.

If students are ever to remove these things from the realm of *science legend* and placed in the realm of *explanatory science*, they must first have some understanding of the forces that shape those objects.

Lab 4.2.3 and 4.2.4 begin to deal with the attractive force of gravity, in contrast with the outward electrical force carried by light.

Gravity and tidal forces shape the rings of Saturn, the moons of Jupiter, and the volcanoes on semi-liquid satellites of Jupiter.

Pedagogical Issues the objects which the students are preparing to study are shaped by two forces: gravity, and the electrical force (carried by light). This exercise provides an opportunity for students to apply their knowledge of gravitational attraction to a new phenomenon.

Science Issues

Tidal forces are in fact quite difficult to describe in detail. They are, however, very important in a large number of phenomena.

Tidal forces are believed to transfer enough energy in some cases to keep some planets molten.

The Learning Activity

This activity is primarily a pencil and paper activity. Some additional reading may be helpful.

Before the activity:
Predict: the tides observed during a full moon, a gibbous moon, and a new moon.
Explain: the prediction

There is no experiment as such to do this time.

The idea that the moon's mass can affect the behaviour of every drop of water in the ocean is a little uncanny.

It is almost certain that students will have stories of other things that tidal forces might affect, such as:

deliver dates of babies
uncontrolled behaviour in prisons and mental institutions
car crashes
unusual animal behaviour

Equipment, Preparation and Resources

Pencil, and the lab manual

Categories:
Knowledge:
Inquiry:
Communication:
Applications, Extensions:

Assessment and Evaluation

Lab 3.4: The Sun and Earth's Gravitational Effects Upon the Moon

Pedagogical Issues

This exercise is based upon a conflict between two opposing influences: the attractive force of gravity, and the scattering electrical force (light).

Because this exercise is strongly related to student schemata, students should not find this concept difficult to deal with.

Science Issues

The scattering force of light would be nearly the same intensity on the moon and on the earth, since both are about the same distance from the sun. However, the attractive gravitational forces on the earth and on the moon are quite different.

A gas molecule struck by light and given energy will acquire high velocity. The necessary escape velocity from earth is about 11 km/s. A gas particle traveling slower than that will not leave earth. Escape velocity from the moon is only 1.3 m/s. Sunlight is able to provide the escape velocity to molecules of gas upon the moon, but not to gas upon near the earth.

The end result is that earth has captured most of the available gas in the vicinity of its orbit, perhaps even the atmosphere of Mars...

The Learning Activity

Again, this is primarily a pencil and paper exercise.

The activity has two parts, each followed by a discussion.
Predict: the effect that light energy will have upon the moon's atmosphere.
Explain: using appropriate representations.

In the second exercise:
Predict: whether Mercury, Venus, or Mars would have an atmosphere that contained water.
Explain: using appropriate representations.

Equipment, Preparation and Resources

An encyclopedia with data on the planets

Lab Manual

Categories:	Assessment and Evaluation
Knowledge:	Quality of explanation
Inquiry:	
Communication:	
Applications, Extensions:	Quality of prediction in regards to new situations

Lab 3.5: Observing the Moons of Jupiter

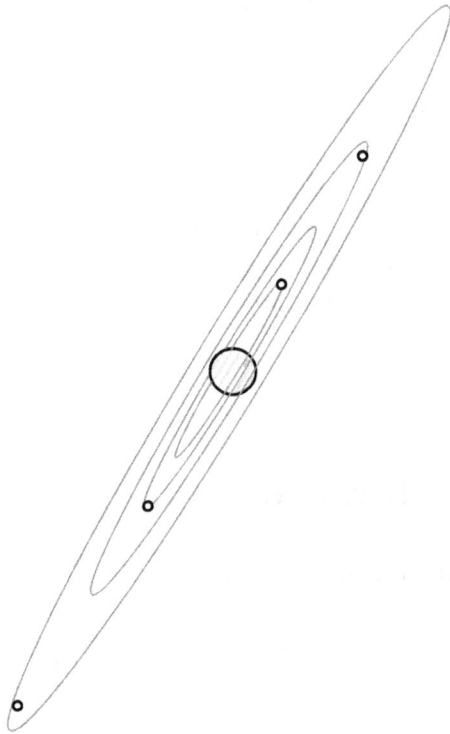

Pedagogical Issues There is no model, no computer animation so exciting and beautiful as the sight of the moons of Jupiter on a clear night.

To capture students' imagination and sense of beauty, there are few sights so lovely as this one.

While you're at it, locate the other major planets. Occasionally Mars, Venus, or Saturn are visible at the same time.

Science Issues Jupiter is a beautiful model of a solar system; in fact, Jupiter's mass is within an order of magnitude of starting its own thermonuclear reaction. When our sun and planets formed from the original disk of matter, had the amount of matter been slightly greater, we might be living in a binary star system, one small solar system orbiting within a larger one.

With binoculars, the moons are clearly visible, but only as bright points of light around the disk of Jupiter. With a terrestrial refractor telescope, about 70 - 100 mm, the radius of the moons' orbits is quite dramatic.

If you do have access to a larger astronomical telescope, perhaps a 150 - 200 mm reflector, the disk of Jupiter is quite clear, with the colour and brightness of each moon being distinguishable.

If viewing is repeated over several hours, the inner moons can be seen to be moving. Over several days, the relative motions of all of the Jupiter's moons can be seen.

The Learning Activity

One evening's viewing with binoculars and one or two larger telescopes is sufficient for this lab. Several nights would be better if someone in the class has access to the telescopes.

Drawings of Jupiter and any other objects observed are necessary, both for records, and for disciplined viewing.

Before the experiment
Predict: which moon will revolve around Jupiter with the greatest speed.
Explain: why you believe your prediction, using diagrams, paragraphs, and the concepts and theories in learned in this course.

After the experiment
Observe: and make records of your observations
Explain: your observations, especially any discrepancies between them and your predictions.

Equipment, Preparation and Resources

Several pairs of binoculars, 50×, with the largest objective lenses you can get.

One or two telescopes. Terrestrial scopes will present better views than most binoculars, but reflecting astronomical telescopes are better.

Categories:	**Assessment and Evaluation**
Knowledge and Understanding:	Explanation of the motion of Jupiter's moons
Thinking and Inquiry:	Quality of prediction :: Explanation
Communication:	Quality of written answers, including diagrams
Applications / Connections:

Project 4.1: Exploring the Rest of the Solar System

Pedagogical Issues   This report is intended to be a pretty standard library / internet research project, with one exception. *Can a student explain how the four fundamental forces shaped the objects in the solar system?*

Since a great deal of effort has been spent upon the four forces, this project offers an opportunity for the teacher to assess how well students can explain phenomena (the planets etc) by appealing to gravity, electromagnetism (light), and the nuclear forces.

Learning skills are emphasized here. Is the student capable of organizing resources and time to complete a project of this sort? Organization skills, reading and note-taking skills, the ability to work both alone and cooperatively are all tested in a project of this kind.

Science Issues   Opportunities exist here to review and apply some knowledge from the chemistry unit. The composition of the many moons within the solar system varies dramatically, from the rocky silicates of the moon to sulfur, methane, slush, and more. Basic description of these substances is within the reach of some students.

The Rest of the Solar System

Science and
Pedagogy

The Learning Activity

This project is intended to require a period of five days, assuming the text and computer resources are available. More time could be provided if the project include one or two weekends as well.

Day 1   Choose an object in the solar system to study. Prepare a list of resources, then locate them: at least two books, two internet sites, and one other.

Day 2   Read resources and make notes. Organize the notes into a rough draft.

Day 3   Write the report. 600 - 1200 words (2 - 4 pages) is not outrageous for a content-based report such as this.

Day 4   Proofread and correct the draft

Day 5   Write the final report:
1 pg   Cover
1 pg   Photo or image
2-4 pg   Body of report
1 pg   Bibliography

Equipment, Preparation and Resources

Access to the internet lab over a period of 2 days
10 - 20 reference books on the solar system

Categories:	Assessment and Evaluation
Knowledge and Understanding:	Depth of factual knowledge displayed
Thinking and Inquiry:	Quality of explanation using the four forces
Communication:	Clarity of writing and argument
Applications / Connections:	Relation of space missions that provided the data used

Project 5.1: Exploring the Rest of the Universe

Pedagogical Issues This project provides a second opportunity to do a research project. The structure of the project is quite similar to the last one, but the subject matter occurs over a much larger scale of space and time.

Organization skills, reading and note-taking skills, and ability to work both with and without the assistance of a group are tested in this project.

Students who found the last project challenging have a second opportunity to master the necessary skills. Students who were quite successful last time will gain confidence and skill in rapidly completing this project.

Science Issues There is simply too much stuff in this unit to teach in any other way but through research projects. A list of some of the possible topics is included in the lab manual.

The Learning Activity

This project is intended to require a period of five days, assuming the text and computer resources are available. More time could be provided if the project include one or two weekends as well.

Day 1   Choose an object in the solar system to study. Prepare a list of resources, then locate them: at least two books, two internet sites, and one other.

Day 2   Read resources and make notes. Organize the notes into a rough draft.

Day 3   Write the report. 600 - 1200 words (2 - 4 pages) is not outrageous for a content-based report such as this.

Day 4   Proofread and correct the draft

Day 5   Write the final report:
   1 pg   Cover
   1 pg   Photo or image
   2-4 pg   Body of report
   1 pg   Bibliography

Equipment, Preparation and Resources

Access to the Internet Lab for one or two days.

A set of approximately 20 reference books.

Categories:	Assessment and Evaluation
Knowledge and Understanding:	Depth of factual knowledge displayed
Thinking and Inquiry:	Quality of explanation using the four forces
Communication:	Clarity of writing and argument
Applications / Connections:	Relation of space missions that provided the data used

2 Explaining the Solar System

Knowledge and Understanding

Three theories are emphasized in this unit. You are probably already familiar with the cell theory. In addition, you will learn new ideas about cell division. Finally, you will learn how our ideas about DNA allow us to explain how living things reproduce. Additional concepts will be introduced as needed.

We will work with pictures and models to illustrate how reproduction occurs.

Knowledge and understanding are probed at regular intervals in the Grade Nine Daily quizzes. Study these as you go through the exercises, so that you can do your best when they are assigned.

Inquiry and Thinking

As often as possible, we will use the PEOE cycle for most labs and activities. You are expected to frame a question, provide your best prediction, and explain your thinking, using both sentences and diagrams. In other exercises, you will make clones of a plant, a simple worm, and yeast. In other exercises, you will use paper models to predict how genetic damage can occur.

At the end of the unit, you will be given a five day independent project. The project will demonstrate your ability to conduct your own investigation.

Communication

The quality of your arguments is the most important aspect of communication in this chapter. Your arguments consist of sentences, organized into paragraphs, and supported by diagrams or other representations.

Each sentence should be clear and to the point. You will find it best to limit your sentences to two concepts linked together to make a reasonable claim. If you need to relate more than two concepts, add a new sentence.

Applications, Connections and Extensions

Every exercise in this book is designed to support you as you learn appropriate theories and apply them to problems. In the labs, you demonstrate your understanding of a theory only by applying the theory. In the quizzes and projects, you are invited to make further connections and extensions of your learning.

Explaining the Solar System

Introduction: Three Theories to Understand our place in the Solar System

What is it really like "out there"? Where did we come from? Are we alone? How did it all begin? How will it all end? Human beings have asked these questions from the very beginning of our existence, and will continue to ask them into the distant future. Some of these questions may never be answered. At the end of this course, you will have some degree of understanding of the various ways that people have sought to answer these questions, and how we think of them today. This unit addresses three main ideas:

1. **There are only Four Forces.** In all of the material universe, four forces connect everything that we can perceive. These forces are the same everywhere: gravity near a black hole is basically the same as gravity on earth, only more intense. We are a part of the universe, so those four forces are present in our bodies and our neighbourhoods as well as the distant galaxies.

    1. **Gravity: a very weak force of attraction between any two things with mass.** Gravitation is the weakest of the four forces. The presence of a piece of matter in space appears to distort the very space and time around it. The result is that gravity appears to affect not only matter, but energy as well. Gravity extends, as far as we know, throughout the entire universe, all space and all time, with the possible exception of the vicinity of the Big Bang.

    2. **Weak Nuclear Force: a force that influences radioactivity inside the nucleus.** You will probably never experience this force in any direct way. It operates only during very small distances and times, and only on fleeting particles inside the nucleus.

    3. **Electric Force: a strong force between any two things with charge. Light is the carrier of the electric force.** Consider two electrons, both negatively charged, that repel each other. How does one electron affect the other electron? We have come to understand that light always begins at one charged particle, and ends at another charged particle somewhere else. It is light that carries the force of electric repulsion from one electron to another. Everything that you see with your eyes is the result of some distant electron losing energy, and your eye receiving that energy. Light is the carrier of that influence.

    4. **Strong Nuclear Force: an extremely strong force that holds nuclei together.** In a previous unit, you learned that all chemical interactions are the result of electrical forces. The nuclear force is 10 000 000 times stronger than the electric force. Thus, nuclear reactions are about ten million times more energetic than chemical reactions. This force is responsible for the nuclear reactions that occur in the centre of the sun, releasing all of the solar energy that we receive each day.

Sun, moon and stars, day and night, summer and winter, the passage of years, the stuff you are made from, the planets in our solar system, absolutely everything involves these four forces.

Some physicists are entertaining the possibility of one or two additional forces. At this time, these forces are not well understood. Stay tuned...

2. **There is only One Earth** upon which we observe the universe. Because the earth we stand upon is spinning, tilting, and revolving around the sun as we look at the rest of the universe, our observations are made difficult.

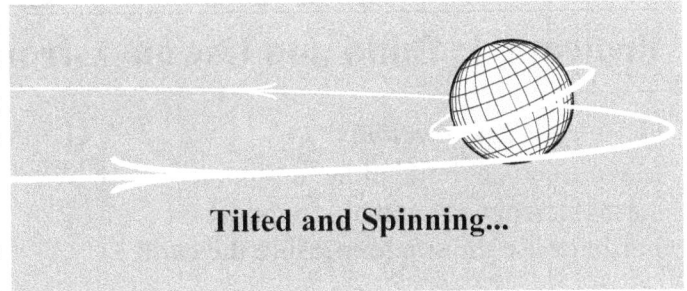

Tilted and Spinning...

Some of this course will help us to make sense of the apparent motions of the sun, the moon, the planets and the stars as we spin, tilt and revolve with our earth. One of the most humbling things about observing the night sky is the size of it all, and how small we are against it.

Our study of space must always reflect the finiteness of our own planet, and the precious ways in which the four forces have shaped our planet to support our lives.

3. **Where do we go from here?** For some of us, our journey as human beings might be a physical voyage to the moon or other planets. Others may study and work to take a journey of exploration, using telescopes, space probes, the Internet and other technological devices. For all of us, a journey of understanding is always possible. We will study the structure of the observed universe, from the most local to the most remote, and from the present to the distant past.

The image at right shows a tiny slice of the sky, about as big as one kernel of popping corn held at arms' length. Each fuzz patch is one galaxy, and each galaxy contains perhaps ten billion stars like our sun.

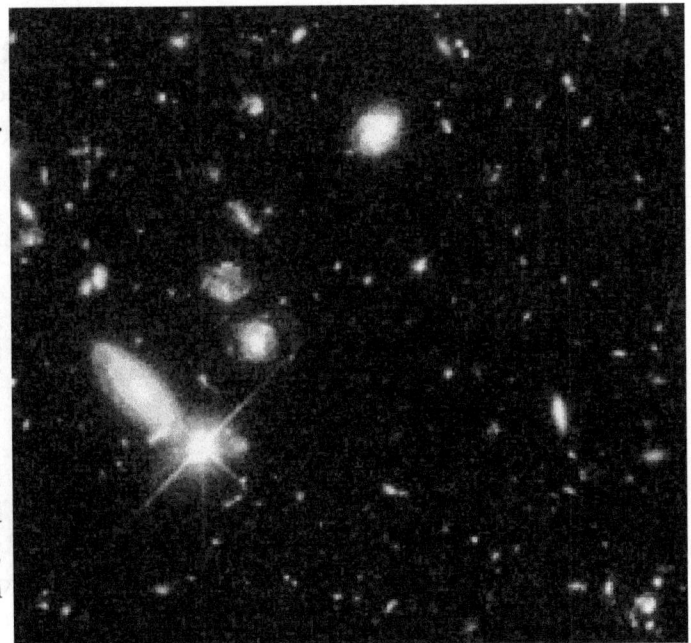

In all of the exercises in this book, the question must be answered in *complete sentences*. One sentence is one thought. A single word is simply not enough.

Project 1.1: Build and Use an Astronomical Instrument

0 *Project Instructions* Make a choice: build either a pinhole solar telescope, a sundial, a support for Earth, or use the sun to measure the earth. From instructions on pp. 34-37, or from another source, draw designs for your project, including: *materials*: cardboard, wood, metal, plastic, *dimensions* of all parts in cm, and *processes*: glue, scissors, saw, drill, etc., involved in use or construction.	**0 My Plan and Outline** When and how I will acquire materials, tools, etc. Date: / 4
1 Begin to build your instrument. Consult with teachers or other sources regarding the best use of available resources, tools, etc.	**1 Rough notes and plans here.** Write enough that your teacher can understand what to expect in your final report. Date: / 4
2 Complete basic construction of the instrument. It should be solidly constructed, not flimsy or messy.	**2 Rough notes and plans here.** Write enough that your teacher can understand what to expect in your one page report. Date: / 4

Project 1.1: Build and Use an Astronomical Instrument

3   Calibrate your instrument.
How is it to be set up?
Is direction important?
Must it be level?

Where must any reference markings be placed? In the sundial, where should the hour markings go?

3   Rough notes and plans here. Write enough that your teacher can understand what to expect in your one page report.

Date: / 4

4   Finish your instrument.

Paint, glaze, enamel, fabric, polish?

Colour? Should it look antique or modern?

4   Rough notes and plans here. Write enough that your teacher can understand what to expect in your one page report.

Date: / 4

5   Demonstrate your instrument.

Supply a two page report:

Page 1: How was the instrument constructed?

Page 2: How should the instrument be used?

Submit your instrument, your two page report, and attach this page.

5   Rough notes and plans here. Write enough that your teacher can understand what to expect in your one page report.

Date: / 4

Project 1.2: Building a Pinhole Solar Observer

MATERIALS / TOOLS
- Cardboard tube, 1.2 m long
- Small mirror
- White plastic shopping bag
- Aluminum foil
- Small cardboard box, or more tube
- Duct tape, glue
- Hand saw, keyhole saw, drill

INSTRUCTIONS
1. Cut the tube to a length of about 1.2 m.
2. Cut an opening in the tube.
3. Cut a ring off the tube, 2 cm wide.
4. Cut out a piece of the ring. Squeeze the ring to make it smaller.
5. Cover the ring with a piece of white plastic grocery bag to make a screen.
6. Push the screen up the tube as shown
7. Place a mirror in the opening, at 45°.
8. Hold mirror in place with a ball of plasticine or duct tape.
9. Cut holes in a cardboard box to fit the tube as shown. Tape it securely to keep out the light. You look down through the box at the mirror to see the screen.
10. Seal up the back of the tube to keep out the light.
11. Put aluminum foil over the front of tube, and tape it securely. Poke a very small pinhole hole in the aluminum foil.

USING THE TELESCOPE
1. Point the telescope at the sun. Look into the cardboard box. You will see a small disc of light: an image of the sun.
2. A larger pinhole produces a brighter but blurrier image. A smaller pinhole produces a sharper but dimmer image.

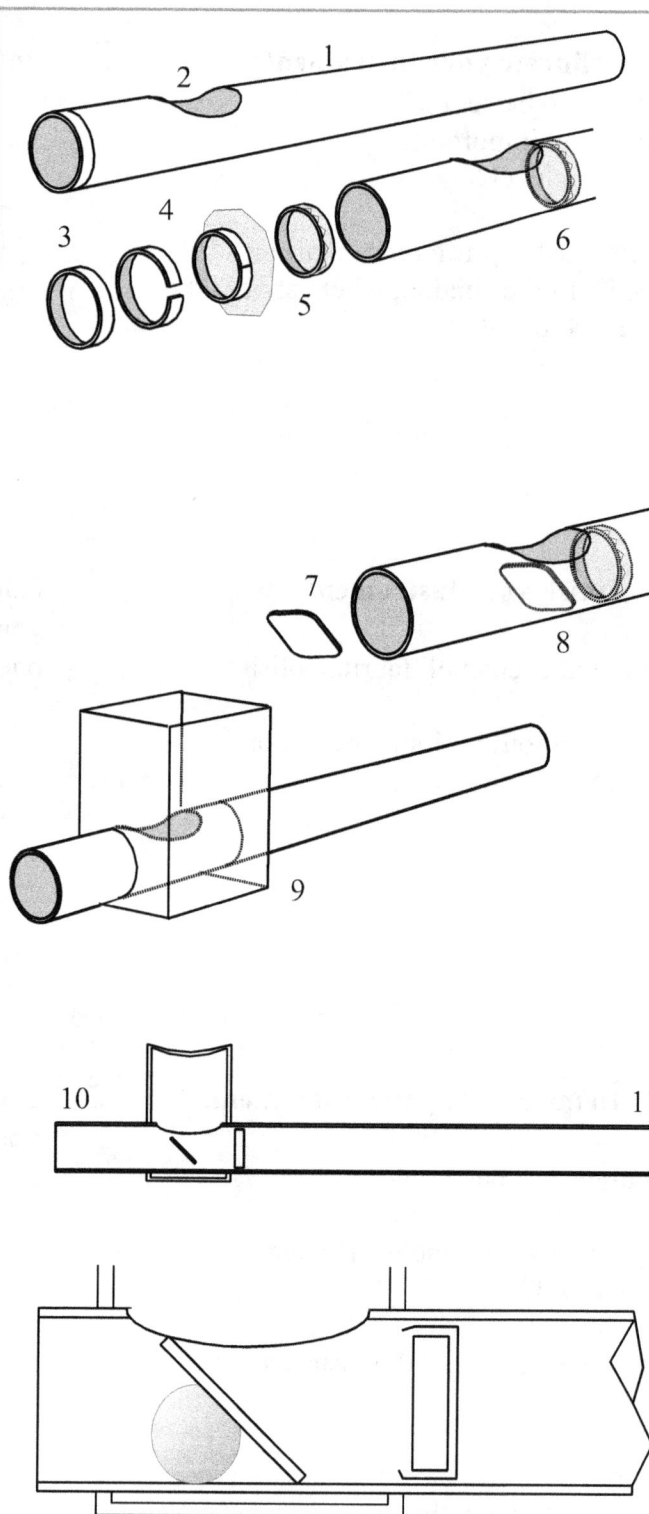

Project 1.3: Using the Sun to Measure the Earth

INSTRUCTIONS

1. Use the Internet and an atlas to search for a school in a city which is at least 500 km due south of your school.

2. Contact the school via email, to find out if they want to participate. Determine the exact distance between you and them.

3. Send them these instructions. Either scan this page and attach it to an email, or fax it.

4. To make a measurement of the angle of the sun: suspend a metre stick from a string as shown. Place a second metre stick along a perfectly North South line on a level surface. Hold the suspended metre stick so that the ends just touch. Measure the shadow of the stick.

5. On a chosen day at 12:00 noon (11:00 Daylight Saving Time) students at both schools measure the length of the shadows cast by their metre sticks, and record them. Exchange the information.

6. What is the angle of elevation of the sun at each school? What is the difference Δ between the angles?

7. The angles may be calculated using some simple trigonometry. The difference in the two angles Δ is equal to the angle shown in the arc on the diagram at right.

8. From Δ and the distance between schools, find out how far it is around the world. Then, find the diameter of the earth.

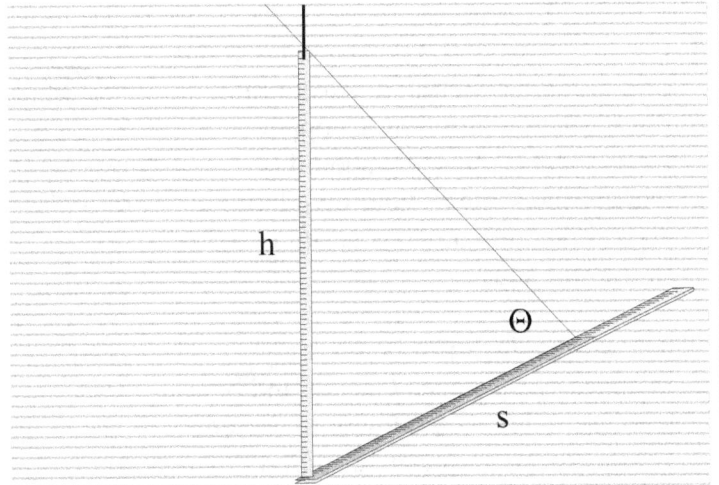

$$\text{Tan } \theta = h \div s$$
$$\theta = \text{Tan}^{-1}(h \div s)$$

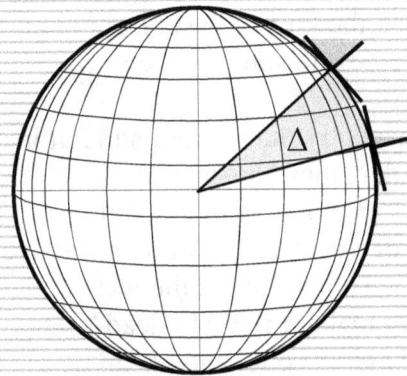

9. If there is 18° difference in the angles of elevation of the sun, then you must be 18° apart on the globe.

10. If 18° corresponds to 500 km, then what is the distance around the whole world (360°)?

11. Now that you know the circumference of the earth, it is easy to find the diameter by the formula:

$$c = \pi d$$

Project 1.4: Building a Sundial

INSTRUCTIONS

1. Cut the knife edge for your sundial out of a durable material, at least 3 mm thick, such as aluminum, copper or brass sheet. The angle should be exactly the same as your own latitude, so that the knife edge is at right angles to the sun's rays.

2. The knife edge must be securely fastened to the base. Copper and brass can be easily soldered with tin based solder. Aluminum can be soldered with silver solder, and a special flux.

3. To calibrate your sundial, set it upon a dead level surface, and point the knife edge due south. Mark the edge of the shadow of the knife edge at each hour of one day, from sunrise to sunset.

4. To permanently mount your sundial in your garden, make a pedestal out of concrete. (Fill a precast concrete flower pot with cement for an attractive pedestal) Tap or vibrate the wet cement to make it settle to dead level, and the base of your sundial can be aligned with the concrete. If you perform this operation at exactly noon, you can turn the sundial so the shadow is thinnest. When the concrete hardens, the sundial will be dead level, and due south.

5. Because the earth is tilted, in summer the sun will be 23.5° higher than average; in winter, the sun's rays will be 23.5° lower than average. What two days of the year would be best for calibrating and mounting your sundial? Explain.

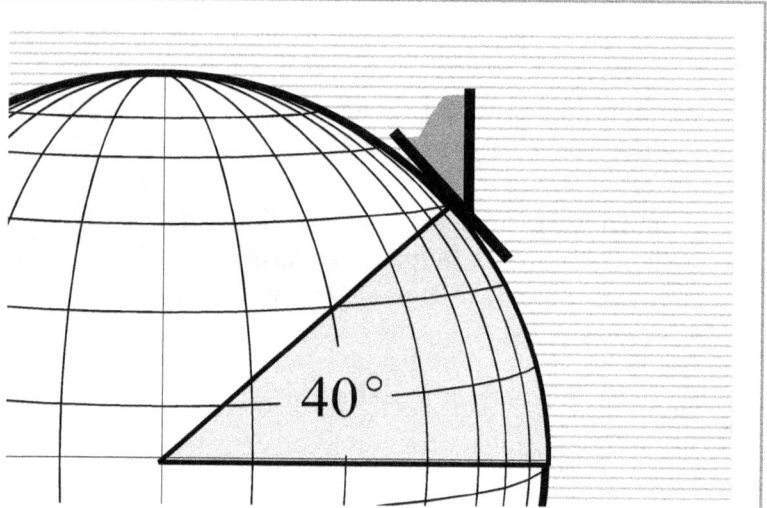

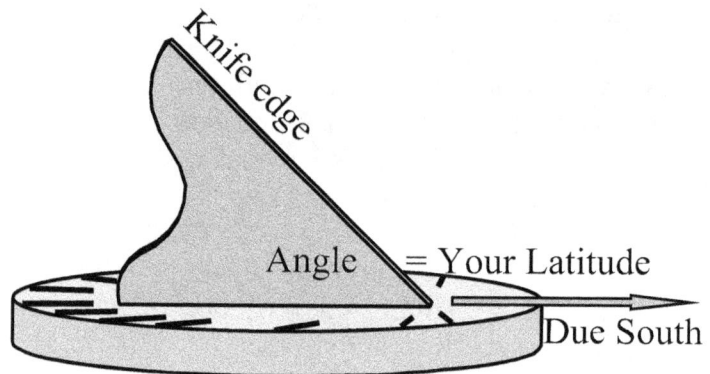

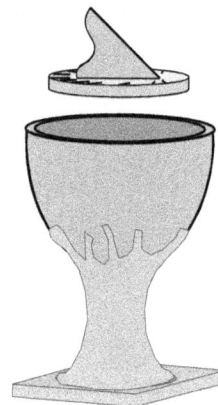

40°

Knife edge

Angle = Your Latitude

Due South

Project 1.5: Building a Support for the Earth

Why do we have seasons? Why does the night sky change with the seasons? What is visible in the night sky at different times of the year?

MATERIALS/TOOLS
Plexiglass, glue, a globe, metal rod
Coping saw, handsaw, drill.

INSTRUCTIONS

1. Measure the diameter of your globe. Let that diameter be x. Round x off to a nice number. The dimensions of the parts A and B are given in the diagram at right.

2. Mark up the materials before cutting. Check the dimensions. Part B should fit from corner to corner on Part A.

3. Make two identical Parts A. One of them (the top) has a hole diameter x cut into it. The other part (the bottom) has no hole cut into it.

4. Make one part B. Cut it into two pieces lengthwise as shown. Cut two notches, the width of the material and half the width of the material. When you place the two pieces at right angles to each other, they will fit together to create the base of the support. The top of the support is then fastened to the base.

5. The top of the support should then be marked with the seasons. The sun is at the location indicated by the corner at each season. The night sky changes with every season. The earth's seasons change because of the 23° tilt of the earth.

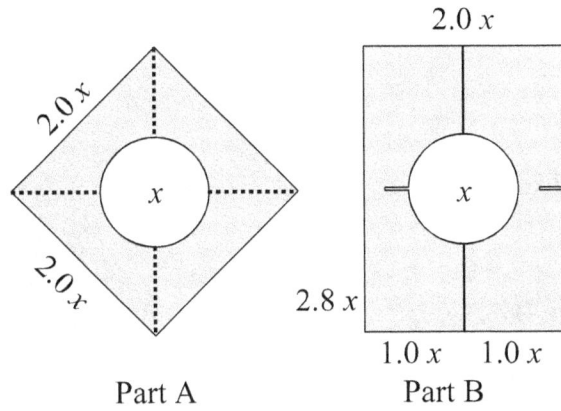

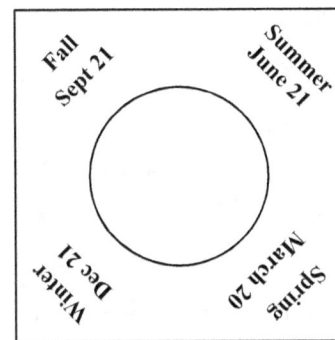

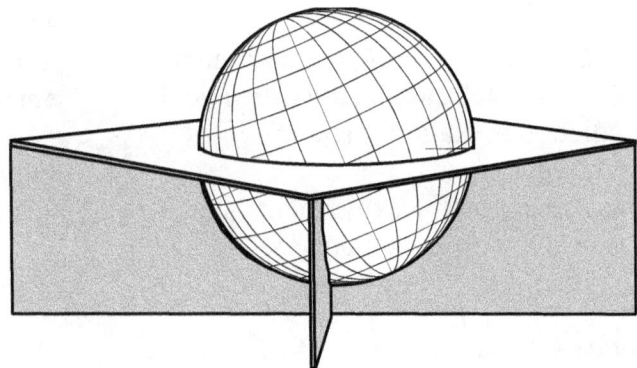

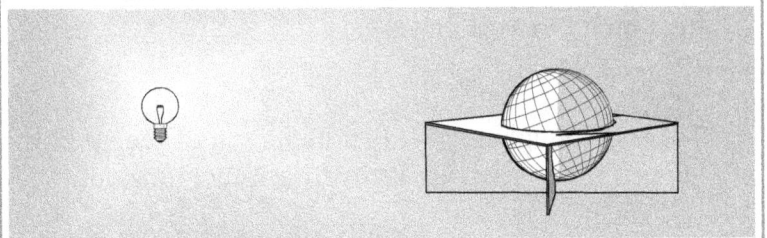

Part A

Part B

Lab 2.1: What Makes a Star Work?

Do you Remember? **List the Four Forces**

1. _____

2. _____

3. _____

4. _____

What's The Question? *What is a star? How does it work?*

What Are We Doing?

1. Inflate a balloon about one-third full. This will be a model of a star.

2. The *skin* of the balloon is like the *inward gravitational force* that holds a star together.

3. The moving *air particles* in the balloon are like the *light* that keeps the sun inflated against the crushing force of gravity.

4. Change the conditions of your model "star" according to the instructions on the opposite page.

5. *Predict* what you think will happen to the star in each case, and *Explain* your prediction in diagrams and in paragraphs.
You can't really do the experiments, but you can check with an authority...

What Are We Thinking About?

1. The more mass in a body, the stronger the gravitational attraction. With enough mass, gravity can create pressures and temperatures millions of times greater than earth's.

2. About 75% of the universe is hydrogen. Although hydrogen nuclei are small, they are positive and repel each other very strongly.

3. Two hydrogen nuclei cannot collide unless the pressure and temperature are about a million times greater than we experience on earth.

4. When four hydrogen nuclei fuse in a nuclear reaction, they produce one helium nucleus, two anti-electrons, and a huge amount of light energy, about ten million times more energy than you got by burning the same amount of hydrogen in a chemical reaction.

5. Light pushes on the particles in the star, holding them up against the crushing force of gravity.

Questions For Later...

1. The sun has been shining with its present brightness for about five billion years. What caused it to light up in the first place?

2. There is enough nuclear fuel in the sun to keep it burning for another five billion years. What will happen to the sun after its hydrogen fuel runs out?

Observing the Sun

Focus Question: Write the question that you are trying to answer.

1 Predict: If you add more mass (hydrogen gas) to your model star (*i.e.,* inflate your balloon some more) ...

1. What happens to the gravitational force?

2. What must happen to the interior pressures and temperatures? How does this affect the nuclear fusion reaction inside the star?

3. What must happen to brightness of the star?

2 Explain your predictions

3 Predict: If the hydrogen fuel gets completely converted to helium, the nuclear reactions will stop. Most of the mass of the star is still there, in the form of the element helium. Because it takes about 100 000 years for the light to get from the core of the sun out to the surface, the core does not immediately cool off.
1. Does the gravity change during this period?

2. Does the star expand or collapse?

3. What happens to the temperature and pressure?

4. It is possible for the temperature and pressure to increase so much that the helium nuclei begin to fuse together to make carbon, nitrogen, oxygen, and even more energy.

5. What will happen to the brightness of the star?

4 Explain your predictions.

Lab 2.2: Observing Sunrise, Sunset and Elevation

What's The Question? We have all seen sunsets, sunrises, and the noonday sun.
Why does the sun undergo its apparent motion?
What can we learn about the earth as we watch the sun?

What Are We Doing?

1. Choose a place close to your home to make an unobstructed observation of the sun from sunrise to sunset.

2. Find due south, east and west from that place, and draw in some landmarks (towers, trees, chimneys etc.) on the diagram page 43.

3. Choose one day of the week (e.g. Thursday). Observe the *location* and *time* of sunrise and sunset on that day each week. Record them on page 43 as in the chart below ◇1.

4. On two different days, 4 weeks apart, observe the sun each hour during the day, recording the position, date and time on the diagram on page 43.

5. On the calendars on the following pages, record times of sunrise and sunset to the nearest minute each day.

6. Record the length of the day and night in hours and minutes on each day of the calendar.

Directly Overhead

1. If your observations are made carefully over a long period of time, you will have an invaluable record of the apparent motion of the sun. On what day of the year is the length of the day equal to the length of the night?

2. On what day of the year is the length of the day changing the most quickly?

Name:
Date:

Observing the Sun

Monthly Solar Observation Calendar 1

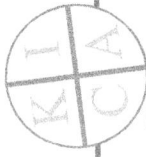

September or February

Sunday	Monday	Tuesday	Wednesday	Thursday	Friday	Saturday
Example Sunrise 6:51 Sunset 19:36 Day 13:15 Night 10:45						

www.rosslattner.ca

Observing the Sun

Name:

Date:

Monthly Solar Observation Calendar 2

October or March

Sunday	Monday	Tuesday	Wednesday	Thursday	Friday	Saturday
Example Sunrise 6:51 Sunset 19:36 Day 13:15 Night 10:45						

Observing the Sun

Name:

Date:

Solar Observation Calendar Diagram

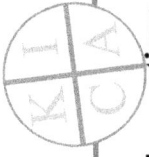

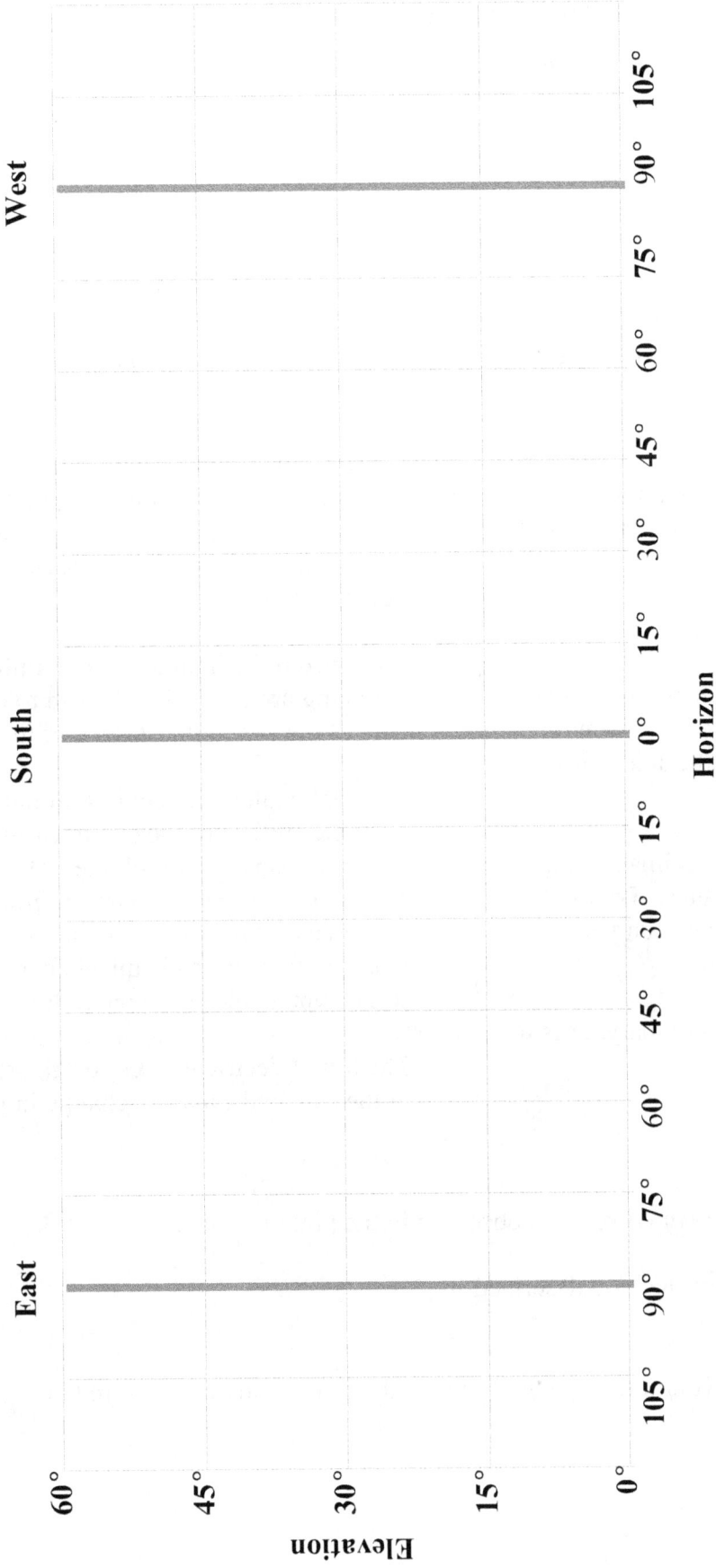

All Observations

Plot all of your observations on this diagram.

© *Ross Lattner Publishing.*

www.rosslattner.ca

Lab 2.3: Observing the Sun with a Pinhole Telescope

Do you Remember? **List the Four Forces**

1. _____
2. _____
3. _____
4. _____

What's The Question? How does the telescope work?
What does the sun look like when observed with a simple pinhole telescope? Is it stationary or is it moving?

What Are We Doing?

1. Some students have built pinhole telescopes to observe the sun. Prior to its use, *Predict* the size and position of the sun's image in your telescope, and *explain* your prediction.

2. Using masking tape, a chair, books, etc, point the pinhole camera directly at the sun so that an image of the sun appears on the screen.

3. Make a sketch of the sun's image, and observe it for a few minutes. Try to *explain* any difference between your prediction and observation.

4. Is the image of the sun stationary, or is it moving?

What Are We Thinking About?

1. All of the light from the sun originated in the core of the sun in a nuclear reaction in which four hydrogen nuclei were fused together by the strong nuclear force.

2. It has taken the light about one million years of bouncing around inside the sun to travel from the core of the sun to the surface of the sun.

3. It has taken the light about 8 minutes to travel from the surface of the sun to earth through the vacuum of space. It took the light about 300 billionths of a second to travel from the pinhole to the screen. It took the light about one billionth of a second to cause a chemical change in your eye, so that you could sense the light.

4. The light (electrical force) originated in the core of the sun, and caused a change in your body!!

Questions For Later...

1. Does the sun appear to be moving when observed in the pinhole camera?

2. What is causing the effect that you described in (1)?

3. How could you modify your telescope to eliminate the effect you described in (1)?

Observing the Sun

Focus Question: Write the question that you are trying to answer.

1 **Predict:** What will the image of the sun look like?

2 **Explain** why you believe your prediction.

3 **Observe,** and record your observations here.

4 **Explain** your observations.

Explaining the Solar System

Lab 2.4: How Big is the Sun?

Do you Remember? List the Four Forces

1. _____
2. _____
3. _____
4. _____

What's The Question? The sun is 150 000 000 km away from us. How far is that in metres? If we know how far away the sun is, can we measure its size?

How big is the sun? Can it be measured from the earth?

What Are We Doing? *Predict* the size of the sun, in metres. *Explain* your prediction, giving both your knowledge and your thinking. *Observe* the sun, and calculate its size. *Explain* your new measurement of the sun, including any differences from your prediction.

1. Use the telescope setup you perfected last day.
2. Measure the length of the tube from the pinhole to the screen, in metres.
3. Measure the size of the image of the sun, in metres.
4. Calculate the size of the sun.
5. Compare your measurement with an authority, like your text book, the Internet, or another reference book.

What Are We Thinking About?
1. The light inside the pinhole telescope makes a triangle. The pointy end of the triangle is at the pinhole. The length of the triangle is the length of the telescope. The width of the triangle is the width of the sun's image.

2. The light outside the telescope makes another triangle, exactly the same shape as the first. The pointy end is at the pinhole, the length of the triangle is the distance to the sun. The width of the triangle is the size of the sun.

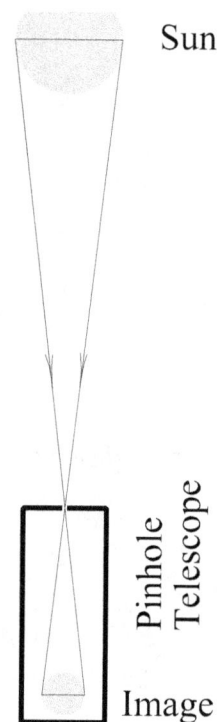

Sun

Pinhole Telescope

Image

Questions For Later...

1. Compare your measurement of the size of the sun with an authority. How big was your error?

2. In order to measure the size of the sun, you had to know the distance to the sun. How could you actually measure the distance to the sun?

Observing the Sun

Name:

Date:

K I
C A

Focus Question: Write the question that you are trying to answer.

1 **Predict** the size of the sun, in metres, without consulting an authority.

2 **Explain** your prediction.

3 **Observe,** and record your observations here.

4 **Explain** your observations.

Lab 3.1: Observing Earth's Largest Satellite

What's The Question? The moon comes up, the moon goes down. Sometimes it's full, sometimes not. Sometimes you see it at night (you expect that), but sometimes the moon is visible in the day.

Why does the moon undergo its apparent motion?

What Are We Doing?

1. Choose a place to make an unobstructed observation of the moon at *the same time each day*. Each person in the class can choose a different time. Make your observations right on the hour.
2. Find due south, due east, and due west from that place, and draw in some landmarks on the diagram below.

3. Observe the location of the moon, and its appearance, three or four times each week for the next four months.

4. **It is important that you make your observations at the same time each day.**

5. Record the *appearance* of the moon on each date, as well as its position. Record the location and date of any observed moonrises and moonsets

6. For practice, on the chart below, draw the moon:

Dec 1	Full	0° E 44° elevation
Dec 3	¾	30° E 40° elevation
Dec 7	½	90° E 8° elevation

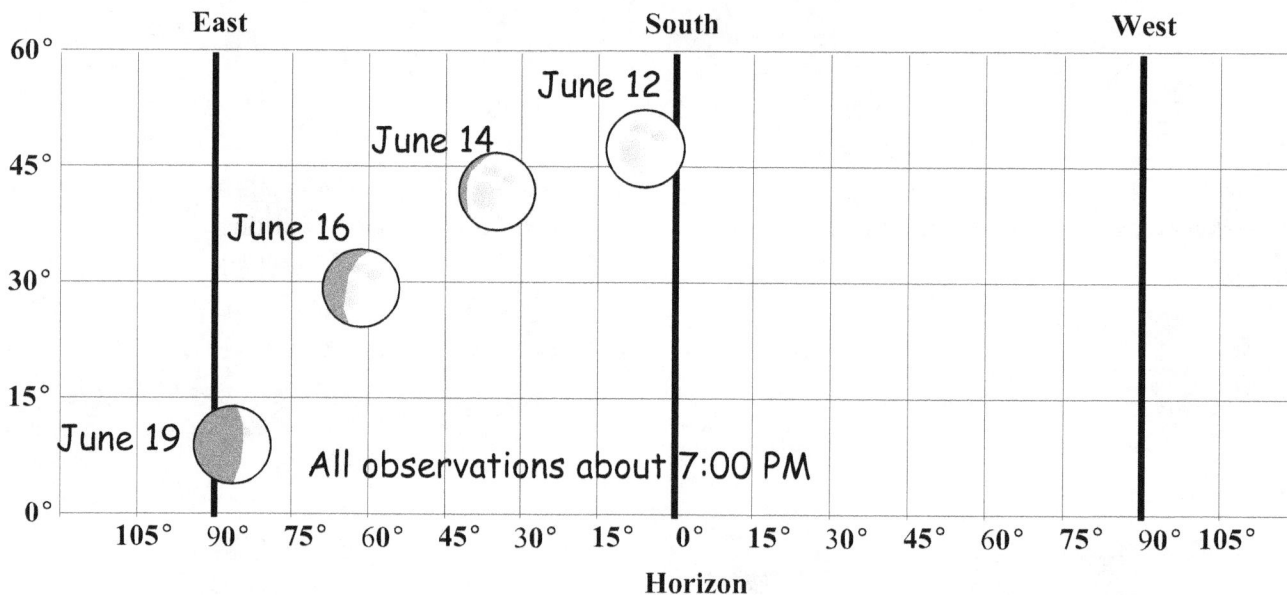

If your observations are made carefully over a long period of time, you will see have an invaluable record of the motion of both the earth and the moon as they dance their monthly rhythm.

What does the moon look like when it rises exactly the same time as the sun sets? Why?

Name:

Date:

The Motion of the Moon

Monthly Lunar Observation Calendar 1

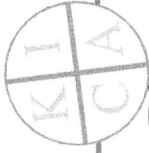

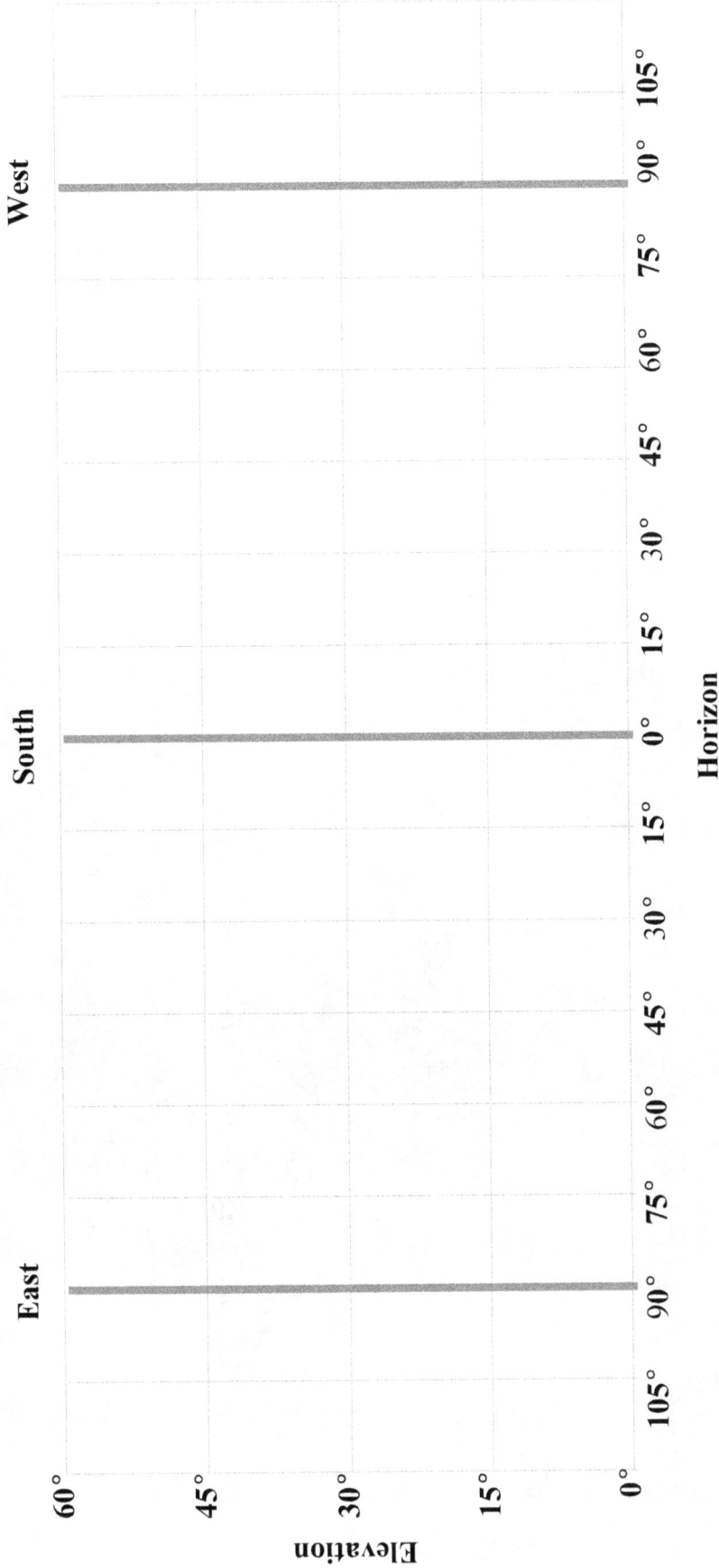

September or February

Start on any date. Make your observations at the same time each day. Draw a *picture* of the moon, in its observed position. Record each *date*. Repeat until the moon disappears at the eastern horizon. Go to next chart.

The Motion of the Moon

Name:

Date:

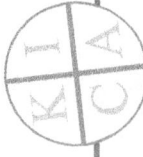

Monthly Lunar Observation Calendar 2

October or March

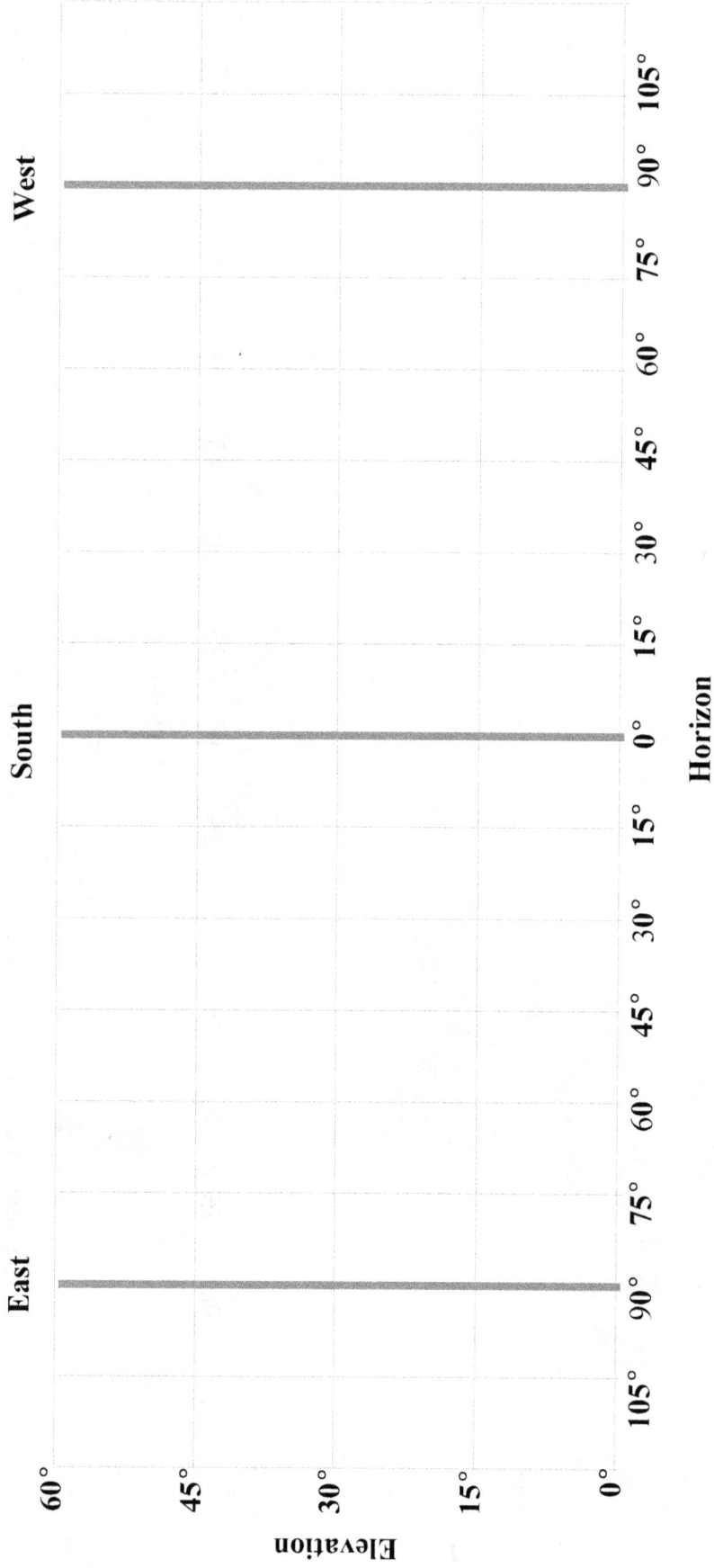

Start recording as soon as the moon appears on the western horizon at your chosen hour of observation. Make your observations at the same time each day. Record the moon's appearance and observed position for two weeks. Draw an arrow on every diagram to indicate the direction to the sun.

The Motion of the Moon

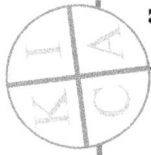

Monthly Lunar Observation Calendar 3

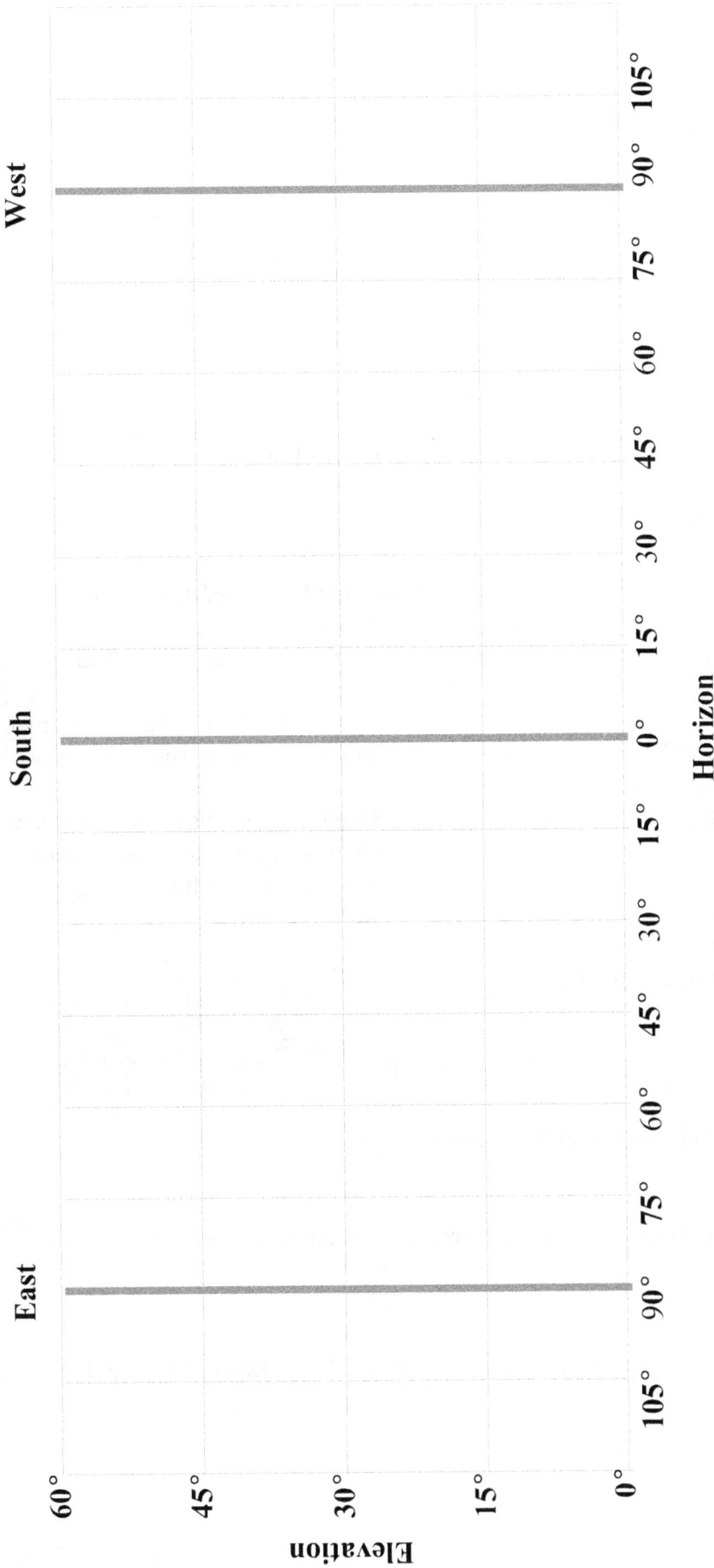

November or April

East   South   West

Elevation

60° 45° 30° 15° 0°

105° 90° 75° 60° 45° 30° 15° 0° 15° 30° 45° 60° 75° 90° 105°

Horizon

Continue to record as before. Where is the sun at your chosen hour of observation of the moon? Draw and date the sun if it is visible. If the sun is not visible, draw an arrow near each lunar diagram, pointing toward the sun.

Continue to draw the moon's appearance and position as before. Start your observations as soon as the moon appears on the western horizon, and continue for two weeks until the moon disappears on the eastern horizon. Draw the appearance and location of the sun at your chosen time of observation. If the sun is not visible, then draw an arrow near the moon, pointing in the direction of the sun.

Lab 3.2: Observing Earth's Smallest Satellites

Do you Remember? **List the Four Forces**

1. _____
2. _____
3. _____
4. _____

What's The Question? Artificial satellites are designed to relay communications, photograph crops, survey the earth, provide navigational signals, and to spy on people. From the earth, a satellite looks like a tiny white speck that moves silently and smoothly without blinking. It might travel across the whole sky in less than a minute
How many artificial satellites pass overhead each night?

What Are We Doing?

1. On any clear night, for the first two or three hours after sunset, artificial satellites are visible directly overhead.

2. Lie down on a dry comfortable place, looking straight up for about 30 min. Make a record of the time and direction of travel of any satellites that you see.

3. Look in the same place at the same time for the satellites on another night. Some navigational satellites are designed to return each night at the same minute.

What Are We Thinking About?

1. A *satellite* is any object in orbit around another, usually larger, body.

2. Artificial satellites have been placed in orbit by humans for particular purposes.

3. Satellites just above the earth's atmosphere are travelling at a speed of about 11 km every second. Those in higher orbits travel more slowly.

Questions For Later...

1. How many satellites did you observe in thirty minutes?

2. At that rate, how many satellites would pass overhead in twenty-four hours?

3. Why do the satellites not fall to earth or fly off into space? What keeps them up there?

The Motion of Satellites

Name:

Date:

K I
C A

Focus Question: Write the question that you are trying to answer.

1 *Predict:* How many satellites do you expect to see in thirty minutes?

2 *Explain* why you believe your prediction.

3 *Observe,* and record your observations here, including apparent brightness, time, location and direction.

4 *Explain* your observations.

Lab 3.3: The Sun and Moon's Gravitational Effects Upon the Earth

Do you Remember? **List the Four Forces**

1. _____
2. _____
3. _____
4. _____

What's The Question?
The earth's gravity holds the moon in orbit. *What effects does the moon's gravity have upon us?*

What Are We Doing?
Consult some reference books or the Internet to help you answer the following questions:

1. How high are the ocean tides, on average?

2 Why are the tides higher on a full moon?

Use your knowledge to answer these questions:

3. The earth's atmosphere can be considered to be an ocean of air. 90% of the earth's atmosphere is below the altitude of 20 km.

4. If the moon's gravity causes tides of water, would it cause tides of atmosphere?

What Are We Thinking About?

1. Suppose that you had a block of cheese on earth, so that the force of gravity on the cheese was one Newton (N).
2. The same block of cheese, on the surface of the moon, would be attracted to the moon's surface at 1/6 N.
3. That same block of cheese, lifted away from the moon, would continue to experience the moon's gravity. At the distance of the earth, the intensity of the moon's gravity is 1/50 000 N.
4. In other words, the block of cheese has a downward force of attraction to the earth of 1N plus an upward attraction to the moon of 0.000 02 N.
5. That tiny force is enough to cause the tides on the earth!

Questions For Later...
1. If you checked the weather records for the past year, would you expect to see clear skies or cloudy skies around the time of a full moon? Explain.

2. Consult your records of the moon's cycles. Can you detect a trend toward clear or cloudy weather during full moons?

Focus Question:

1 *Predict* the tides observed on a *full* moon, a *gibbous* moon, and a *new* moon.

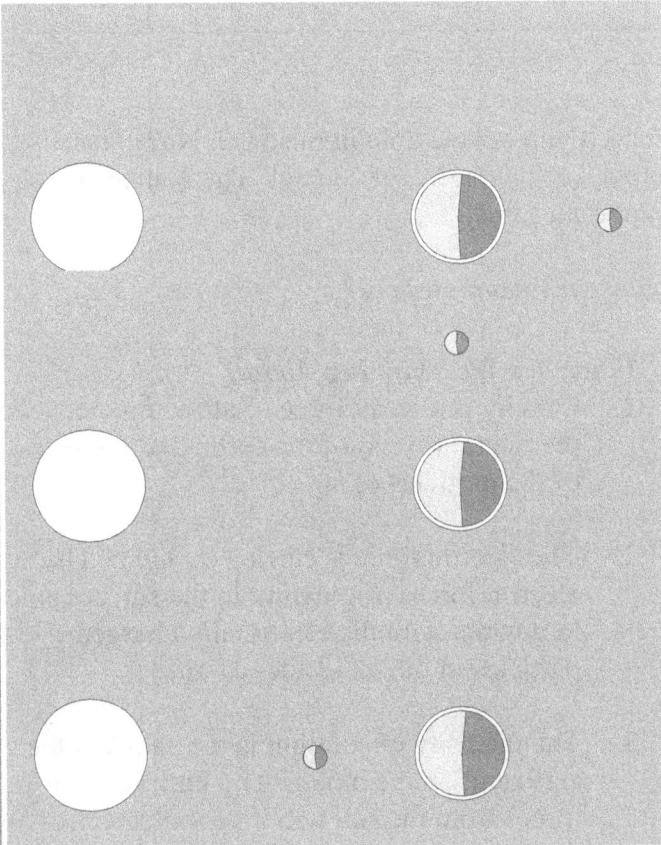

2 Would the tides in the atmosphere occur in the same manner as tides in the water? *Explain* your predictions.

3 *Predict* the atmospheric pressure during periods of high tides, and periods of low tides. Would a full moon tend to cause clear skies (high atmospheric pressure) or cloudy skies (low atmospheric pressure)?

4 *Explain* your predictions.

Lab 3.4: The Sun and Earth's Gravitational Effects Upon the Moon

Do you Remember? **List the Four Forces**

1. _____

2. _____

3. _____

4. _____

What's The Question? Can you keep air in a plastic bag if you poke a hole in the bag? No? Then what keeps the earth's atmosphere from leaking off into space? Gravity, of course! The matter in the earth attracts every molecule of air with a small but important force.

The moon has gravity. *Why doesn't the moon have water, or an atmosphere?*

What Are We Doing?

1. Find the chemical formula for water, and for the molecules that make up the atmosphere.
2. Add up the atomic masses of the atoms in each molecule to find the molecular mass of each molecule.
3. Find the Gravity Escape Quotient for Earth by dividing its *mass* (kg) by its *radius* (m). Then find the Gravity Escape Quotient for the Moon.

Predict the outcome when the electrical force (light) and the gravitational force act on the atmosphere and water on Earth and the Moon *Explain* your prediction.

Observe the earth and moon, and any references that have relevant information.

Explain what you found in your research.

What Are We Thinking About?

1. Gravity is a weak force of attraction between the matter (mass) of the earth, and the mass of the particles of air.

2. The electric force is carried by light. The electric forces originating in the sun are able to interact with the electrically charged particles in the molecules of air.

3. The intensity of sunlight is the same on the earth and on the moon. The light (electric force) from the sun will give molecules of air the same energy.

4. It takes more energy to escape the earth than it does to escape the moon.

Questions For Later...

1. Mars is just a little larger than our moon. Would you expect Mars to have an atmosphere? Explain your reasoning.

The Motion of Satellites

Focus Question: Write the question that you are trying to answer.

1 **Predict** the effect that light energy (the carrier of the electric force) will have on the moon's atmosphere.

Where do you think the Moon's atmosphere and water ended up?

2 **Explain** (use the diagrams below).

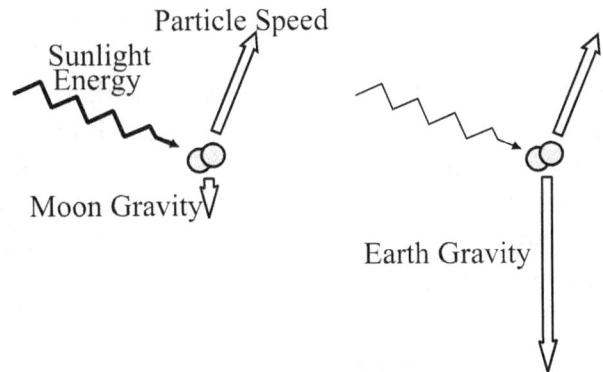

Particle Speed

Sunlight Energy

Moon Gravity

Earth Gravity

3 **Predict** whether Mercury, Venus or Mars would have an atmosphere or water.

If there ever was water on Mercury, Venus or Mars where are the water particles now?

4 **Explain** your findings.

Lab 3.5: Observing the Moons of Jupiter

Do you Remember? **List the Four Forces**

1. _____

2. _____

3. _____

4. _____

What's The Question? Jupiter is a very large planet. Its mass and gravitational field are so large that as Jupiter orbits the sun, it makes the sun wobble a bit. If the Earth and Jupiter are in the right position, you can find Jupiter just after sunset, or just before sunrise, along the *ecliptic*. It looks like a bright star, one of the first to appear in the evening sky. The moons of Jupiter are easily visible with a pair of binoculars. You can see them moving from one hour to the next. Without binoculars, a person with sharp eyes might occasionally be able to make out moons as a tiny streak of light.

What Are We Doing?

1. Get a few people together to observe the sky.

2. Find Jupiter. (Don't know where to look? Ask your teacher, or consult the Internet or a reference book. The computer program "Starry Night" is an easy way to find things in the sky).

3. Use a pair of binoculars to view Jupiter. Make simple sketches of Jupiter plus its moons to show their relative positions.

Predict which moon would move the fastest. *Explain* why you believe your prediction.
Observe Jupiter and its moons over a period of several hours or days.
Explain your findings, using diagrams and paragraphs.

What Are We Thinking About?

1. How does Jupiter's radius and mass compare to Earth's?

2. How does Jupiter's gravity compare to Earth's?

3. Some scientists believe that Jupiter is nearly large enough to become a star. If its mass was just 20 times larger, Jupiter's gravitational force might have crushed the core to the enormous temperatures and pressures needed to start a nuclear reaction. In that case, the moons of Jupiter would have been a miniature solar system!

4. Most of Jupiter's mass is gaseous molecules. From where did they come?

Questions For Later...

1. Do the moons of Jupiter ever move into Jupiter's shadow? That would be a *lunar eclipse*. Draw a diagram to show how such a thing could occur.

2. Do the moons of Jupiter ever cast a shadow on Jupiter's surface? That would be a *solar eclipse*. Add to your drawing to show how such a thing could occur.

The Motion of Satellites

Focus Question:

1 *Predict* which moon will revolve around Jupiter with the greatest speed.

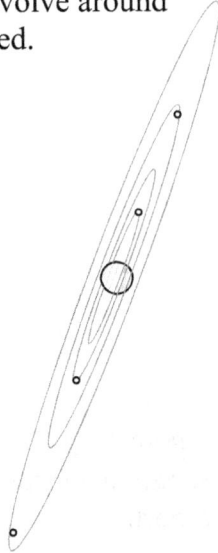

2 *Explain* why you believe your prediction. Use diagrams, paragraphs, and the concepts and theories that you have learned.

3 *Observe,* and record your observations here.

4 *Explain* your observations.

Project 4.1: Exploring the Rest of the Solar System

0 ***Project Instructions*** Choose an object in the solar system, and make a thorough study of its:
- Mass, radius, rotation, orbit, day, year, gravity and other features
- Composition, atmosphere, surface, temperature, pressure, and so on
- Satellites, rings, gas clouds, etc.
- Space probes, scientific missions, earth bound telescope explorations, and Hubble projects that have contributed to our knowledge.

0 My Plan and Outline.

Date:

1 In addition to your text, you must consult at least two books; two Internet sites; and one other resource of your choice. Prepare your library and Internet search strategy.

Execute your strategy, locating if possible more than you think you will need to use. Make proper bibliographic notes on each resource.

Write a one page report of your findings.

1 Rough notes and plans here. Write enough that your teacher can understand what to expect in your one page report.

Date:

2 Read your resources, and make notes.

Compare and contrast resources, especially between newer and older sources.

Organize your notes into a rough draft.

Include discussion of the four forces, and how they have shaped the planet.

2 How do you plan to make notes as your read? File cards? Sheets of loose leaf? Computer? Be sure to include page number, book etc to every note!!

Date:

Project 4.1: Exploring the Rest of the Solar System   Name:

3 Write your report. The whole report should be 600 - 1200 words, *plus* bibliography and cover. In addition, you must include at least one full page picture, diagram, or photo of your subject.	**3** Pay attention to organization. Make each paragraph count! Date:
4 Give your report to someone else to read. At the same time, you read someone else's report. Make comments upon spelling, grammar, clarity of writing, and so on. Return the papers. Polish up and revise your own report.	**4** What do you need to fix in your report? Don't say "nothing". You can always improve. Date:
5 Submit a finished report: Cover one page Photo or picture at least one page Body of report two to four pages Bibliography one page Attach this page for teacher assessment and evaluation	**5** After the project: what would you do differently next time? Date:

The Five Day Project

Project 5.1: Exploring the Rest of the Universe

0 *Project Instructions*

This assignment is the final project for the unit. Choose an *object* or *process* or *phenomenon* that takes place outside our own solar system. In every case, you must describe it thoroughly, but you must not stop at that. You must explain *why* that object, process, or phenomenon exists as it does.

To do so, you must investigate the role of gravity, the role of the electric force (light) and perhaps the role of the nuclear force.

Some possible topics might be:
- Why is the night sky black? Why is that important to your understanding of our universe?
- What is the "Big Bang" theory?
- What is the "background microwave radiation?" Why is that important to your understanding of our universe?
- What happened in the first three minutes of the universe?
- What are "The Pleiades?" What is happening there?
- Where is the Orion Nebula? What is it?
- Where is the Andromeda Nebula? What is going on there?
- Where is the Trifid Nebula? What is going on in there?
- Where is the Horsehead Nebula? What does it tell us about our galaxy?
- What is a spiral galaxy?
- What is an elliptical galaxy?
- What is a "black hole," and where are they found?
- What is a "red giant?" How did it come to be?
- What is a "white dwarf?" How did it come to be?
- What is a "nova?" How do they come about?
- What is a "supernova?" What makes them occur?
- Can galaxies collide? What might happen if they do?
- What is a "quasar?"
- What is a "neutron star?"
- What is a "pulsar?"
- What is a "globular cluster," and where is it found?
- The universe was originally 75% hydrogen, and 25% helium. Where did the other elements come from?

You may wish to investigate other topics. *Seek prior approval from your teacher.*

Project 5.1: Exploring the Rest of the Universe

Name:

Star Map: Where in the sky can we find the object you are studying? Draw a map of the sky big enough to show at least three constellations. Provide the location of the object you are studying, and some other instructions about finding it.

Project 5.1: Exploring the Rest of the Universe   Name:

0 *Project Instructions* Choose an object in the universe, and make a thorough study of its properties. 1. What is this object? 2. What is going on inside it? How does it work? Include the four forces! 3. What is its relation to the rest of the universe. 4. Discuss scientific projects that have contributed to our knowledge.	**0** My Plan and Outline Date:
1 In addition to your text, you must consult at least two books, two Internet sites, and one other resource of your choice. Prepare your library and internet search strategy. Execute your strategy, locating if possible more than you think you will need to use. Make proper bibliographic notes on each resource. Write a one page report of your findings.	**1** Rough notes and plans here. Write enough that your teacher can understand what to expect in your one page report. Date:
2 Read your resources, and make notes. Compare and contrast resources, especially between newer and older sources. Organize your notes into a rough draft. Include discussion of the four forces and how they have shaped the planet.	**2** How do you plan to make notes as your read? File cards? Sheets of loose leaf? Computer? Be sure to include page number, book, etc. to every note!! Date:

Project 5.1: Exploring the Rest of the Universe   Name:

3 Write your report.

The whole report should be 600 - 1200 words, *plus* bibliography and cover.

In addition, you must include at least one full page picture, diagram, or photo of your subject.

3 Pay attention to organization. Make each paragraph count!

Date:

4 Give your report to someone else to read. At the same time, you read someone else's report.

Make comments upon spelling, grammar, clarity of writing, and so on.

Return the papers.

Polish up and revise your own report.

4 What do you need to fix in your report? Don't say "nothing". You can always improve.

Date:

5 Submit a finished report:

Cover   one page
Photo or picture   at least one page
Body of report   two to four pages
Bibliography   one page

Attach this page for teacher assessment and evaluation.

5 After the project: What would you do differently next time?

Date:

The Hazards	The Safe Way
In this column is a list of lab safety issues that you will face in this course.	**Read this column to find out how to safely handle the laboratory problem.**
Eye Injury is possible from flying fragments of metal, glass or chemicals; from heat or flames; from caustic solutions such as acids or bases.	*Always wear safety glasses* in the laboratory. Never take your glasses off, even if you have finished your experiment. Other students may not have finished their lab work. The safety glass symbol indicates exercises in which safety glasses *must* be worn.
Crowding, Pushing and Horseplay increase the likelihood of a serious injury.	*Attend to your work.* Stay at the station you were assigned, so that there is room to work safely. If your teacher finds that your behaviour is a safety hazard, he or she may remove you from the lab. There is no place for behaviours which place others at risk of injury. Not at school, not at home and not at work.
Disorganized and Dirty Working Conditions are a hazard wherever they are found.	*Keep Lab Area Clean.* Clean and put away unused equipment. Tell your teacher about chipped, cracked, damaged or broken equipment. Do not leave anything on the floor, the desktop, the sink, or the cupboards that is not supposed to be there.
Broken Glass happens even to careful scientists.	*Do Not Touch* broken glass with your hands. Tell your teacher. When instructed to do so, use a broom to sweep the glass into a dustpan. Dispose of the broken glass in the special container provided. Do not leave it in the regular wastebasket: it could seriously injure a custodian.
Liquid Spills may consist of water, but they may also contain acids, bases, or toxic chemicals. You may not be able to tell the difference.	*Tell your teacher* about any spills immediately. Do not attempt to clean up without teacher instruction. Only if the teacher decides it's safe, use a cloth or paper towels to soak up excess liquid. Wipe the area clean with a damp cloth. Rinse the cloth frequently in fresh water. Wash your hands afterwards.
Solid Spills may consist of highly reactive chemicals. You may not know the specific hazards.	*Tell Your Teacher* about the spill, whether or not you caused it. Your teacher will instruct you on the safe way to handle the problem. In any case, the spill must be cleaned up promptly.

Appendix: Laboratory Safety

Open Flames are a frequent hazard. The Bunsen burner is the most likely safety hazard.	***Review Safe Handling of a Bunsen Burner*** with your teacher. Be prepared to show how to light, operate and extinguish the burner at any time. Do not attempt to ignite pens, papers, rulers or other things. That kind of behaviour will certainly result in your being put out of the lab.
Fire. Any liquid solid or gaseous fuel burning where you do not want it to burn is a fire.	***Tell the teacher immediately!*** Do not attempt to extinguish the fire with your hands, books, paper towels etc. Do not panic. Move away from the hazard. ***Your teacher is the best judge of the appropriate course of action.***
Hot Metal or Glass cause more burns than any other hazard. There is usually no visible indication that they are hot. Glass in particular causes small, deep burns.	***Let Hot Objects Cool for 10 - 15 Minutes*** before handling. Place all hot objects on a heat resistant pad. You and your partner will know where they are. Approach hot objects cautiously. Touch them at the coolest point first (the base of the retort rod, the bottom of the Bunsen burner or hot plate, the thumb screw of the iron ring). Use dry, not damp, paper towels to handle hot objects.
Hot Liquids such as boiling water or hot oil spread and splash rapidly. They also cling to skin and clothes.	***Let Hot Liquids Cool for 10 - 15 Minutes*** before handling. Do not heat liquids in closed containers. Use hot plates rather than shaky retort rod assemblies. Do not heat more liquid than you need.
Obstructed Passageways prevent you from moving out the way of a spill or a fire.	***Stand at Your Lab Station.*** Do not bring chairs or stools over to sit down. Your chair will prevent others from moving away from a spill or a fire.
Long Hair or Loose Clothing is more likely to become involved in your equipment. It can cause spills and breakage, or catch fire.	***Tie Back Long Hair; Secure Loose Clothing.*** Outerwear in particular must be avoided in the lab situation. Jackets, sweat suits, hoods, etc are too large and awkward for the lab situation. They are also frequently made of materials that are flammable and can melt and stick to the skin in a fire. Avoid using laquer based hair sprays. A curly head of hair with hair spray can burn up completely in seconds.
Unauthorized Experiments can have unintended results.	***Stick to the plan.*** Read instructions very carefully the night before the lab. Ask questions. Do not try experiments "just to see what happens." The dangers are too great.